Hidden Futures

Hidden Futures

Death and Immortality in Ancient Egypt, Anatolia, the Classical, Biblical and Arabic-Islamic World

Edd. J.M. Bremer, Th.P.J. van den Hout, R. Peters

Amsterdam University Press

This publication has been made possible by the generous support of NWO.

Illustration on cover: Attic funerary vase of ± 440 AC. W. Riezler, *Weissgründige Attische Lekythen,* München 1914, II 44. A.
Riezler's comment: "In seinem Kahn ist Charon, ein wüster Geselle; an dem Strand des Acheron gefahren, hält er den Kahn mit seiner Ruderstange fest und wartet auf die Tote, die verhüllt steht und zu zögern scheint. Hermes aber streckt gebietend die Hand nach ihr aus, um sie dem Charon zu übergeben. Ein deutliches Zeichen der Örtlichkeit sind die vielen Eidola, die Seelchen, die Charon umflattern..." (I 11-12)

Cover design: Erik Cox, Den Haag

Lay-out: A-zet, Leiden

ISBN 90 5356 078 5

Table of contents

Archaeological

Introduction

Few moments in man's life are so certain and yet so fraught with fear of the unknown as death. It is a challenge to which every civilization responds in its own fashion. Attempts at dealing with this fear range from complete acceptance or resignation in the sight of death to suppression and denial of its inevitable coming. In some civilizations, the promise of a paradisiac future can turn life in this world into only a temporary stay, a relatively unpleasant preliminary to a far better eternity. In others, any form of afterlife is denied, and the unknown is transformed into the certainty of non-existence, leaving man with no other goal to pursue than the *hic et nunc* of this world. It is self-evident that these various views determine to a large extent the ceremonies used to part with the body of the deceased and the literary genre used to express grief about the death of friends and relatives or to console the bereaved.

There are probably as many views on death as there are civilizations; even within one such civilization, views may shift from one century to another, as is exemplified in ancient Greece. This has been observed for our own era as well: in our culture death once seemed to be banned to sterile hospital beds and funeral homes, while nowadays it tends to become more 'social' and comes into the open again. On the whole, a renewed interest in death can be noticed these days, and the present book is an expression of that trend.

This volume of essays grew out of a symposium held in December 1992 at the University of Amsterdam and organized by the Institute for Mediterranean Studies. Most of the papers read there have been included in this book, supplemented by some others, so as to cover as much as possible the broad geographical spectrum of cultures in the area. Although representing only a small part of the many cultures that since prehistory occupied the borders of the Mediterranean, the contributions in the present volume do reflect the wide variety of ideas on Death and Immortality to be found in that region. And, of course, they have common traits as well. In order to bring both differences and similarities to the fore, we have chosen a thematic set-up for this book rather than a chronological or geographical one. Furthermore, the subject index will enable the reader to find his way through the various essays.

The first four contributions offer a general description of ideas and beliefs on death and the afterlife in Ancient Egypt, Hittite Anatolia, Biblical Israel and Classical Greece. As these civilizations have long since disappeared, the term "general" might be somewhat misleading: they can only draw a general picture as far as literary and/or archaeological sources allow us insight into the various cultures. This is, in different ways, especially true for the Hittite empire and Ancient Israel. The main source for the latter, the Bible, is not a systematic treatise on death and the afterlife, but a corpus of a mainly narrative character that went through a long process of canonization. Moreover, the Israelites do not seem to have been preoccupied with death and man's destiny afterwards as, for instance, the Egyptians and other cultures in the Ancient Near East with their multiple divine powers involved. For them nature as a whole, including death, obeyed the will and the word of their one god. So, in search for such views, the modern scholar has to scrape together his evidence scattered throughout this work. In his contribution, Nico van Uchelen discusses the material thus won from different angles: the role death plays in man's dependent relation to God, the way this is expressed in literary form, especially in Psalms and Job, and how death was dealt with on a personal-emotional level. Here we meet Jacob mourning his son Joseph, believed to have been killed by a wild beast, or David tearing his clothes over Saul and Jonathan. It is on this level that we learn most about funeral customs and beliefs in an afterlife.

The restrictions the Hittitologist sees himself confronted with are of a different kind: although here, too, there is no systematic Hittite treatment of the subject, the rich and varied material at our disposal, especially the detailed description of a royal funerary ritual, enables us to obtain a fairly clear picture of the views held on death and the afterlife in Hittite society. Our textual sources, however, predominantly reflect life in the court circles: what the "average" Hittite's thoughts on these matters were remains almost completely in the dark. Only archaeology sheds some faint light on this area. After discussing some general conceptions emanating from the texts and after having compared them with the archaeological material at our disposal, the royal ritual is discussed in detail by Theo van den Hout. With the help of typological comparisons with funerary customs in Imperial Rome, Renaissance Europe and other cultures, he tries to come to a better understanding of some of the many symbolic actions carried out during the 14 days this death rite lasted. One of these, the funerary meal served to the deceased, is echoed in the burial customs of Ancient Latium, as dealt with by Marijke Gnade in the archaeological section (see below).

Quite different is the situation for the Egyptologist: the Egyptians are renowned for their concern with death and immortality, and their material and literary culture, abundantly preserved as it is, seems to be pervaded with funerary associations. In his contribution, therefore, Henk Milde has imposed upon himself chronological restrictions and describes the conceptions of the New Kingdom (ca. 1550-1080 BC), mainly as they have come down to us in the famous Books of the Dead. Different, too, is the attitude towards the dead body. In the Bible the body comes from dust and will return to dust; the Hittite royal corpse is burnt, and the soul transcends to heaven; the Egyptian ceremony, however, is completely aimed at recomposition of the body, the person and his or her body being identical. The many excellently preserved mummies amply attest to this belief. Milde describes the long and difficult journey the deceased had to undertake before being able to enter the realm of the dead under the rule of Osiris. Here he was judged, his purity tested. Surprising, however, is the fact that the deceased, whether king or ordinary human being, is depicted doing heavy agricultural labor in the so-called farmer's paradise.

Finally, the approach of trying to elucidate certain features in a specific culture through cross-cultural comparisons as seen in the contribution on Hittite Anatolia is used in the article by Jan Bremmer. Focusing on "The soul, death and the afterlife in early and classical Greece", his discussion of the various terms for "soul" in early Greek is seen from the perspective of "primitive soul belief" in other cultures, and the changing attitude towards death from Homer to the classical era is compared to the development of ideas on this subject in Western Europe since the early Middle Ages. These changes in attitude regarding death also led to a revaluation of the Greek concept of the Netherworld: the gloomy and macabre image of the Homeric realm of the dead gradually takes on paradisiac features.

Homer, of course, returns in other contributions. A possible Near-Eastern forerunner to Homer's account of Odysseus' encounter with the Netherworld can be found in the Hittite texts and is discussed by Van den Hout. In the literary section, Jan Maarten Bremer offers an elaboration and further illustration of some points made by Bremmer. In Greek poetry one finds sublime pathos (Homer) next to eloquent simplicity about death. Epitaphs give witness to various sentiments vis-à-vis the dead and to the attempts to console those staying behind. Here we see admiration for soldiers fallen or love for children taken away early, but mortality as such is accepted with resignation, as a fact of nature. There are, on the other hand, poems which betray a religious expectation of immortality, or even proclaim certainty about it. The mysteries of Demeter and Dionysus have inspired poets to imagine grandiose visions of "happiness hereafter". The theme of consolation will recur in Bartel Poortman's paper in the philosophical section below.

Homer figures prominently in the contribution by Hans Smolenaars on Vergil's Aeneid. His paper concentrates in the first place on the similarities and differences between the two great poets in dealing with "death on the battlefield": so central in the reality of ancient warfare and in epic poetry which comes to terms with it. In the Iliad death is rampant, culminating in Achilles' killing of Hector, but it is the serenity of lamentation, supplication and reconciliation, which gives the poem its final stamp of validity. At the end of Vergil's Aeneid, however, Aeneas refuses Turnus' supplication and finishes him off in a relentless way. Smolenaars tries to explain Vergil's choice for this brutal act in the historical context of Octavianus establishing his power in the years 30-20 BC.

The two remaining contributions in the literary section are closely interrelated. Both Pieter Smoor and Arie Schippers deal with elegiac poetry in the Arab world. The main motifs of Arabic elegiac poetry, as it developed during the first centuries of Islam, were of pre-Islamic origin and derived from the notions that were common among the pagan Arab Bedouins of the Arab peninsula. Smoor examines the development of these motifs, especially the notions of blind fate, of cosmic grief, and of the futility of human action. He also addresses the question of a possible Hellenistic influence on the genre, which seems to be only incidental in character.

The Hebrew poetry that flourished in Muslim Spain during the tenth and eleventh centuries clearly betrays its dependence on the Arabic poetic tradition. Schippers analyses Hebrew Andalusian elegies, compares their motifs and structures with the Arabic genre, and attempts to establish the debt owed by these Hebrew elegiac poets to their Arab colleagues. Although this genre as an expression of grief, common to all mankind, is universal, both Andalusian Hebrew and Arab poets used the same themes and motifs deriving from the pre-Islamic, Arab culture. While emulating their sources, however, they give evidence of originality and ingenuity.

The struggle with certain death and man's attempts to cope with the tantalizing uncertainties that surround his fate thereafter are perhaps most explicitly addressed in the philosophical treatises. The papers of Bartel Poortman and Rudolph Peters deal with examples of different schools of thought clashing over this theme. Greek philosophers, for instance, have reflected upon death from Socrates onwards. Poortman shows how the possibility of personal survival after death was denied by realistic philosophers like Democritus, Aristotle and Epicurus, whereas Plato's metaphysics seem to have been "invented" as a frame to protect the immortality of the soul. Most surprisingly, popularizing treatises combine arguments from both philosophical schools for a shallow rhetoric of consolation.

The dilemma between the realistic and the metaphysic may also be seen in the way Muslim religious scholars in the nineteenth century tried to reconcile modern science and traditional Islamic thoughts. One of them, Husayn al-Jisr (1845-1909), enjoyed some popularity because he tried to show that both were not necessarily in conflict. In this essay Rudolph Peters examines al-Jisr's ideas on death and the afterlife in order to determine to what extent these were influenced by modern scientific notions and, secondly, what was the function of al-Jisr's use of modern science. The conclusion is that he was not a modernist and did not depart at all from the classical dogmas of Islam: he used the results of the natural sciences in a very eclectic manner to bolster these traditional religious truths.

The last paper in this volume is the only strictly archaeological contribution. Taking an Iron Age cinerary urn in the shape of a hut in the Amsterdam archaeological Allard Pierson Museum as her starting point, Marijke Gnade discusses the "Sitz im Leben" of the whole category of these cinerary urns. They all stem from Central Italy and date from the period between the Late Bronze Age and the Early Iron Age (10th to the beginning of the 8th century BC). Taking into account different aspects like the funerary ritual, social ideology, ceramic production and architectural reality of these hut urns as reflected in the archaeological data, she cautiously proposes seeing in the deposition of the urns the expression of social distinction of specific members in a community.

However differently the various cultures deal with death and despite the wide variety of answers to the question of what comes after it, the papers collected in this book show how the cultures represented here give evidence of parallel developments or have sometimes even influenced each other. But, primarily, they attest to the basic and undoubtedly universal human need to gain a degree of certainty about this hidden future.

Having come to the end of this introduction, we want to express our feelings of gratitude to the Nederlandse organisatie voor Wetenschappelijk Onderzoek for their financial support in the production costs of this book, and to the Amsterdam University Press for their willingness to publish it and giving it its final form. Finally, we thank Petra Goedegebuure for her invaluable assistance in the preparation of the manuscripts.

Amsterdam, July 1994

Jan Maarten Bremer
Theo van den Hout
Rudolph Peters

GENERAL

"Going out into the Day"
Ancient Egyptian Beliefs and Practices concerning Death

"To the West, to the West,
you favoured one,
to the beautiful West ..."

Henk Milde

INTRODUCTION

The ancient Egyptians are known for their concern with death and immortality. They were, however, a life-loving people and by no means inclined to contempt for the world. The way tomb-owners had themselves (and their relatives) portrayed betrays a sheer appreciation of human life. All funerary efforts tend to its prolongation in the hereafter, under better circumstances, of course. This expectation is expressed, for instance, in the title of spell 110 of the Book of the Dead (BD) concerning the Offering or *Hotep* Field:

> Beginning of the spells of the Offering Field,
> spells of Going out into the Day, of coming and going in the realm of the dead,
> of entering the Field of Rushes, of staying in the Offering Field,
> the great abode "Mistress of Winds".
> Having power there, being glorious there,
> ploughing there and reaping,
> eating there, drinking there,
> making love there;
> doing everything that used to be done on earth
> by the copyist of the temple of Ptah, Nebseny, lord of reverence,
> engendered by the draughtsman Thenna, justified,
> born to the housewife Mutresti, justified.[1]

The mentioning of the deceased's name, titles and relatives is very important. They perpetuate your personality. You do not have a name, you *are* a name; just as you *are* a body. As your body should be carefully embalmed, so your name

1 E. Naville, *Das aegyptische Todtenbuch der XVIII. bis XX. Dynastie*, 3 vols. (Berlin, 1886), I: Pl. CXXI, 1-3.

should be conscientiously maintained. Otherwise you lose identity. The Egyptians did not strive after reincarnation of the Hindu type: rebirth in another body, *i.e.* as another person. They yearned for resurrection of their own body, because they wanted to remain the same person with the same name. "*Denn die Heilserwartung besteht in nichts anderem und nichts Geringerem als dem simplen und zugleich erhöhten Bewahren der kleinsten, jedoch zugleich wesentlichen Einheit historisch gewachsener Form: einem Bewahren des Einzelnen in seiner einmaligen Prägung*", as Morenz phrased it.[2] In India, the departed are cremated in order to detach the Self completely from the body in view of an entirely new incarnation. The Egyptians, however, took measures to the opposite effect. Hence a text like BD 89: "spell for letting the *b3*-soul rejoin its corpse in the realm of the dead" (see Fig. 1). The prospect of being burned was horrific. Evil-doers were treated that way in the hereafter in order to annihilate them completely.

Fig. 1 B3-soul rejoining the mummy of Nebseny

"A FAIR BURIAL"

The funeral customs vary according to time and wealth. Here we will deal briefly with the conceptions and rituals of the New Kingdom (*c.*1550-1080 BC). Important sources of information are the decorated tombs of well-to-do people. Books of the Dead should be mentioned, too. In principle, anybody could have the disposal of these funerary spells. The book *Amduat* on the other hand, revealing "what is in the netherworld", was confined to royal tombs, at least in this period.[3] As to the Book of the Dead there were no restrictions whatsoever, except for the price of such a papyrus. Most of the owners belong to the social strata of officials and qualified workmen of the royal tombs. Lower classes could not afford to buy a copy.

---

2 S. Morenz, "Ägyptischer Totenglaube im Rahmen der Struktur ägyptischer Religion", *Eranos Jahrbuch* 34 (1965), 427.

3 The only exception is the tomb of Useramun, vizier of queen Hatshepsut (1490-1468 BC).

In the first half of the eighteenth dynasty, the representations of the burial are traditional. Tomb walls are decorated with a rather loose composition of different funeral scenes.[4] All sorts of rituals are depicted, many of them obsolete a long time ago. After Tuthmosis IV (*c.* 1400 BC) we discern a tendency to omit those rituals from the program that were no longer performed. The diversity of the scenes made way for a single and sequential representation of the actual funeral. This is what we generally see in the illustrations of the Book of the Dead, the so-called "vignettes". The oldest copies display just the transportation of the bier.[5] Later manuscripts are provided with a more detailed version of the funeral. The vignette in question belongs to BD 1, an opening spell in many manuscripts. After the initial adoration of Osiris, the text begins:

> Beginning of the spells of going out into the day,
> of praises and glorifications,
> of going out and in, in the realm of the dead,
> to be spoken on the day of the burial of the late N,
> coming in after going out.[6]

Several days may have passed between the day of the burial and the moment of death. On a stela from the time of Tuthmosis III (*c.* 1450 BC) has been written:

> A fair burial comes in peace, when your seventy days are completed in your embalmment-place. You are placed on a bier ..... and drawn by young cattle. May the ways be opened by milk, until you reach the entrance of your chapel. May the children of your children be collected all together, weeping with affection. May your mouth be opened by the lector priest, may you be purified by the *sem*-priest. Horus adjusts for you your mouth, he opens for you your eyes, your ears, your body, your bones, so that all of you is complete. Spells of glorification will be read for you, an offering will be made for you, while your heart is really with you, your heart of your earthly existence. You will come in your former appearance as on the day when you were born.[7]

---

4 So, for instance, in the tomb of Rekhmire from the time of Tuthmosis III (c. 1450 BC): N. de G. Davies, *Paintings from the Tomb of Rekh-mi-Rê' at Thebes*, (New York, 1935), Pl. XXIV. Also in J. Settgast, *Untersuchungen zu altägyptischen Bestattungsdarstellungen*, (Glückstadt, Hamburg, New York, 1963), Taf. 14.

5 P. Barthelmess, *Der Übergang ins Jenseits in den thebanischen Beamtengräbern der Ramessidenzeit*, (Heidelberg, 1992), 157.

6 B. Lüscher, *Totenbuch Spruch 1 nach Quellen des Neuen Reiches*, (Wiesbaden, 1986), 14-17.

7 N. de G. Davies, "Teḥuti: owner of tomb 110 at Thebes" in *Studies presented to F. Ll. Griffith*, (London, 1932), 289; Pl.40.

Fig. 2a Funeral of Nebqed

In Fig. 2 we see the funeral of Nebqed, who died in about 1400 BC. His mummy is attended by the widow beside the bier, which is placed on a funeral ship (Fig. 2a). At stem and stern two wailing women keep watch over the mummy. They represent the goddesses Isis (head) and Nephthys (foot), who bewailed their brother Osiris. The goddesses are also depicted as kites. The shrill shrieks of these birds, imitated by wailing women, is thought to raise him from the dead. Lament is not only an expression of sorrow and grief; it is also a means of resuscitation.[8] Behind the bier with the mummy four men drag a chest. It contains the four canopic jars with the deceased's viscera, removed from his body during the embalmment. A priest in the appearance of the divine embalmer Anubis is looking after it. The mortal remains are followed by male and female mourners. A current lament is, for instance, the motto given above: "To the West, to the West, you favoured one, to the beautiful West ..."[9] Compared with the quiet attitude of the men, the mourning gestures of the women display the usual vehemency. Canopic chest and funeral ship have been put on sledges so that they can be dragged over sand towards the tomb. The chest is dragged by men, the funeral ship by cattle (Fig. 2b), as the stela described. Before the ship a *sem*-priest in a panther-skin libates and censes. Behind the cattle a man sprinkles milk. Originally, pouring water served the facilitation of the dragging. At the same time it was a cultic purification. Or better, the purification served the "opening of the ways"; especially when milk was used. It symbolized purity by its colour and, as a vivifying substance,[10] opened the way to life.

The tomb-chapel is marked by the sign of the West. The mummy is put erect before the entrance, facing the *sem*-priest and the widow kneeling down with grief. Other relatives are not depicted in this manuscript. Above the *sem*-priest we read, in accordance with the text on the stela: "purifying the scribe Nebqed, justified". He is depicted, however, with an adze (𓌻), an instrument for the Opening of the Mouth. This is not in contradiction with the text on the stela,

8 Chr. Seeber, "Klagefrau", *LÄ* III: 444-447. J. Zandee, "Sargtexte, Sprüche 363-366 (Coffin Texts V 23-28)", in *Funerary Symbols and Religion*, ed. J.H. Kamstra, H. Milde, K. Wagtendonk, (Kampen, 1988), 176-177.

9 E. Lüddeckens, *Untersuchungen über religiösen Gehalt, Sprache und Form der ägyptischen Totenklagen*, (Berlin, 1943), 129-130 (Nr. 61, from the tomb of Hormin in Saqqara).

10 *RÄRG*, 459.

Fig. 2b Opening of the Mouth of Nebqed

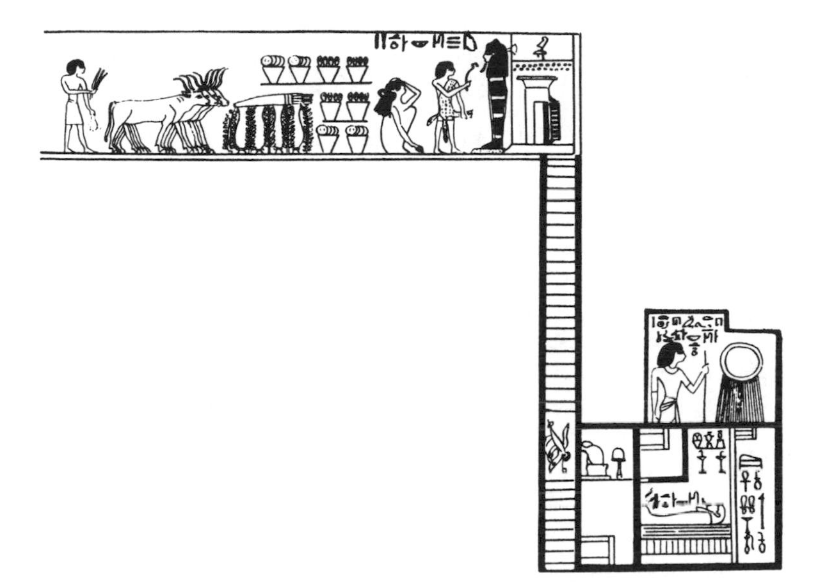

where this ritual was said to be performed by the lector priest. Purification forms part of the entire ceremony. Whenever the lector priest is depicted, we see him with a papyrus in his hands reciting the spells of glorification. The ritual was very complex, various priests being involved.[11] The *sem*-priest not only purifies, libates and censes, he also manipulates instruments. Offering is an important part of the ceremony, too; especially the offering of a heart and a foreleg of a calf. In the papyrus of Hunefer (*c.* 1300 BC) we see two men running away with these pieces, leaving the mutilated calf behind (Fig. 3). Among the instruments belonging to the ritual of the Opening of the Mouth, the foreleg is already depicted in anticipation. In this manuscript we see other priests, too, such as the embalmer, embracing the mummy, and the lector priest (upper left corner).

The entire ceremony aims at the recomposition of the body, which will serve as a home for the deceased's personality. "He will come in his former appearance as on the day when he was born", the stela assured. Hence the emphasis on the heart of the earthly existence. As the centre of all deliberation and affection, it constitutes the deceased's personality, that he eagerly wants to save.

---

11 E. Otto, *Das ägyptische Mundöffnungsritual*, Wiesbaden, 1960; G. Goyon, *Rituels funéraires de l'ancienne Égypte*, (Paris, 1972), 87-182; J. Assmann, "Egyptian Mortuary Liturgies", in *Studies in Egyptology*, ed. Sarah I. Groll, 2 vols., (Jerusalem, 1990), I: 1-45.

Fig. 3 Opening of the Mouth of Hunefer

Then the actual interment follows. The papyrus of Nebqed is unique in showing this (Fig. 2b). Descending the shaft we reach several tomb-chambers. Half-way we meet what has occasionally been called the "external soul":[12] the *b3*-bird, returning to the mummy in order to join it as shown in Fig. 1. Spell 1 of the Book of the Dead ends up with the words:[13]

> O you, who give bread and beer to the excellent *b3*-souls in the House of Osiris,
> may you give bread and beer to my *b3*-soul with you, time and again.
> O you, who open ways and open up roads for the excellent *b3*-souls in the House of Osiris,
> open the ways for me, open up the roads for me, for my *b3*-soul with you,
> so that he may enter vehemently and leave peacefully the House of Osiris.
> Nobody will ward him off or turn him back.
> He enters favoured, he leaves loved.
> His voice is right; what he commands is done in the House of Osiris.
> I have come here and no fault of mine is found:
> "the balance is void of his guilt",
> namely the nobleman and prince, seal-bearer of the King of Lower Egypt, the sole friend, the god's father of the Lord of the Two Lands, Yuya, justified, lord of reverence.

12 G. van der Leeuw, *Phänomenologie der Religion*, 3rd ed., (Tübingen, 1970), 331 (§ 42,4).

13 In the version of Yuya; see E. Naville, *The Funeral Papyrus of Iouiya*, (London, 1908), Pl. II. In the papyrus of Nebqed the spell itself has not been copied.

In accordance with this passage we see the *b3*-bird of Nebqed carrying bread and beer. Another feature is the deceased's unimpeded access to the House of Osiris. In the meantime, however, we have entered the Mountain of the West, descended into the netherworld. In other words, we have crossed the border between this world and the next.[14] There we witness the resurrection of the deceased. It is not without meaning that Nebqed is depicted above the tomb-chamber walking in the light of the sun. He is "going out into the day", according to the legend, free from mummy-wrappings. This is in a nutshell what it is all about in a Book of the Dead. That is why the Egyptians themselves called these papyri "(Books of) Going out into the Day".[15]

THE HOUSE OF OSIRIS

The access to the House of Osiris was dependent on the deceased's morality. Therefore he was interrogated by 42 netherworld judges, holding court in the doorway leading to Osiris. They refuse admission, until they are persuaded by the deceased's declaration of innocence. This trial is symbolically represented by the so-called *psychostasy* (Fig. 4).[16] The heart of the late Hunefer (▽), constituting his personality, is balanced against *ma'at*, the principle of good order, hieroglyphically represented by a feather (𓆄).

Fig. 4 Psychostasy *and introduction to Osiris*

14 J. Zandee, "De reis van een dode", *JEOL* 15 (1957-58), 65-71. A. Piankoff, *The Wandering of the Soul*, Princeton, 1974. Chr. Jacq, *Le voyage dans l'autre monde selon l'Égypte Ancienne*, Paris, 1986.

15 The qualification "*Todtenbuch*" (Book of the Dead) was introduced by R. Lepsius, *Das Todtenbuch der Ägypter* (Leipzig, 1842), 3-4, in substitution for the misleading "Rituel funéraire".

16 Chr. Seeber, *Untersuchungen zur Darstellung des Totengerichts im Alten Ägypten*, München, 1976.

When "the balance is void of his guilt" (BD 1, just cited), the deceased is ushered into the House of Osiris. However, not only the netherworld judges have to consent; the deceased also needs the approval of the different parts of the abode of Osiris which he is about to enter.

> ...
> "I will not let you pass by me", says the threshold of this door, "unless you tell my name."
> "'Ox of Geb' is your name."
> "I will not open to you", says the bolt of this door, "unless you tell my name."
> "'Toe of his mother' is your name."[17]

Finally the deceased has penetrated as far as the door-keeper.

> "I will not announce you", says the door-keeper of this Hall of Justice, "unless you tell my name."
> "'Who knows hearts, who investigates bodies' is your name."
> "To which officiating god shall I announce you?"
> "Tell it to the Dragoman of the Two Lands."
> "Who is the Dragoman of the Two Lands?"
> "He is Thoth."
> "Come!", says Thoth, "what have you come for?"
> "I have come here to be announced."
> "What is your condition?"
> "I am pure from evil, I kept away myself from the quarrels of those on duty, I am not among them."
> "To whom shall I announce you?"
> "You shall announce me to Him whose roof is fire, whose walls are living *uraei*, the floor of whose house is water."
> "Who is he?"
> "He is Osiris."
> "Proceed! Behold, you are announced.
> Your bread is the *wḏ3t*-eye; your beer is the *wḏ3t*-eye; your invocation-offerings on earth are the *wḏ3t*-eye."[18]

Now it becomes clear why the parts of the Hall join in the interrogation. The abode of Osiris is not some chapel or other. It is the realm of the dead itself, ruled by Osiris, founded in the primeval waters, governed by the sun, and bordered by protecting *uraei*, the aggressive cobras. Here the deceased wants to enter; not to

---

17 Budge 264, 11-16.
18 Budge 266, 4 - 267,11.

leave the Hall immediately again with an entrance permit for some other paradise.[19] This *is* paradise, being in the presence of Osiris. But the deceased is not tolerated here, unless he is pure. Hence the interrogation by the 42 judges at the entrance. Hence the interrogation by the "chapel" elements, which can be understood as a mythical reduplication. The vignette in the papyrus of Hunefer illustrates this interpretation of the abode of Osiris. The roof is fiery because of the sundiscs carried by erect *uraei*. The walls consist of the bodies and tails of these cobras.[20] The floor on which the god is enthroned is the primeval water. The blue lotus, symbol of resurrection,[21] springs from it. So it is not a religious building; it is the realm of the dead itself. It has got the shape of a chapel, because it is the holy domain of the netherworld-god Osiris.

Especially since the Ramesside period of the New Kingdom (*c.* 1300 BC) the tomb was related to the netherworld.[22] Entering the tomb-chamber of Sennezem (TT 1), one is confronted first of all with the god Osiris, depicted opposite the entrance.[23] Before Sennezem's mummy was introduced into this domain of Osiris, he was purified by the *sem*-priest during the ceremony of the Opening of the Mouth. So his condition was pure. Once inside looking to the left, *i.e.* to the west, we see the late Sennezem and his wife in adoration before "all the gods of the netherworld". The divinities are seated in a "chapel" similar to the one of Osiris. Osiris himself is depicted among them and so is the sungod Re. Looking in the opposite direction we see a picture of the eastern horizon: the sungod adored by baboons as he rises from an area with lush vegetation. This paradisiac region, sometimes called a farmer's paradise,[24] is well-known from the Book of the Dead, too. It is the vignette of the Offering Field, which belongs to spell 110 mentioned in the Introduction.

---

19 The scribe of the papyrus of Neferubenef is the only one introducing BD 125d with the words: *ḏdt ḫft prt m m3ᶜ-ḫrw m wsḫt nt m3ᶜty* (S. Ratié, *Le papyrus de Neferoubenef (Louvre III 93)*, (Cairo 1968), Pl. XVIII = Naville, *Todtenbuch* II: 310 [Pb]); apparently a misinterpretation.

20 See the snake of BD 87 in the pYuya *e.g.* (Naville, Iouiya, Pl. IX).

21 A. Erman, *Die Religion der Ägypter*, (Berlin, Leipzig, 1934), 62. G.A.D. Tait, "The Egyptian Relief Chalice", *JEA* 49 (1963), 96.

22 *RÄRG*, 258. J. Assmann, "Das Grab mit gewundenem Abstieg. Zum Typenwandel des Privat-Felsgrabes im Neuen Reich", *MDAIK* 40 (1984), 277-290; J. Assmann, "Priorität und Interesse: das Problem der Ramessidischen Beamtengräber", in *Problems and Priorities in Egyptian Archaeology*, ed. J. Assmann, G. Burkard, V. Davies, (London, New York, 1987), 31-41. K.J. Seyfried, "Entwicklung in der Grabarchitektur des Neuen Reiches als eine weitere Quelle für theologische Konzeptionen der Ramessidenzeit", *ibid.*, 219-253.

23 B. Bruyère, *La tombe nº 1 de Sen-nedjem à Deir el Médineh*, (Cairo, 1959), Pl. XIX-XXI, XXIX.

24 H. Kees in *RÄRG*, 161; H. Kees, *Totenglauben und Jenseitsvorstellungen der alten Ägypter*, 3rd ed. (Berlin, 1977), 205. M. Heerma van Voss, "Van ploegen tot aren lezen", in *Beginnen bij de letter Beth*, ed. K.A. Deurloo, F.J. Hoogewoud, (Kampen, 1985), 111.

A FARMER'S PARADISE

The vignette of BD 110 is in fact no more than a map of a typical Egyptian landscape: arable land intersected by canals (Fig. 5). The deceased Nebseny is depicted ploughing and reaping man-sized crops. He can do so without trouble, because there are neither fishes nor snakes in the area.[25] Maybe that is to say that the land to be ploughed is not soaked any longer like the Egyptian soil after inundation, nor too dry as in the Egyptian summer. The dimension of the arable land accounts for a rich harvest: "its extent is the extent of the sky", it says in the curve before the cattle. No wonder then that there is abundance. This is represented by a heron on a perch, . It is somewhat surprising, however, to see how Nebseny gives the abundance away. The "souls of the spirits" benefit from his charity. They manifest their presence by a triple . But Nebseny is not a victim of his generosity. In the next scene we see him on a comfortable chair, his hand stretched out to a well-provided offering-table. "Thousands of all things good and pure for the soul of the writer Nebseny, lord of reverence", the legend assures.

Fig. 5 Offering Field

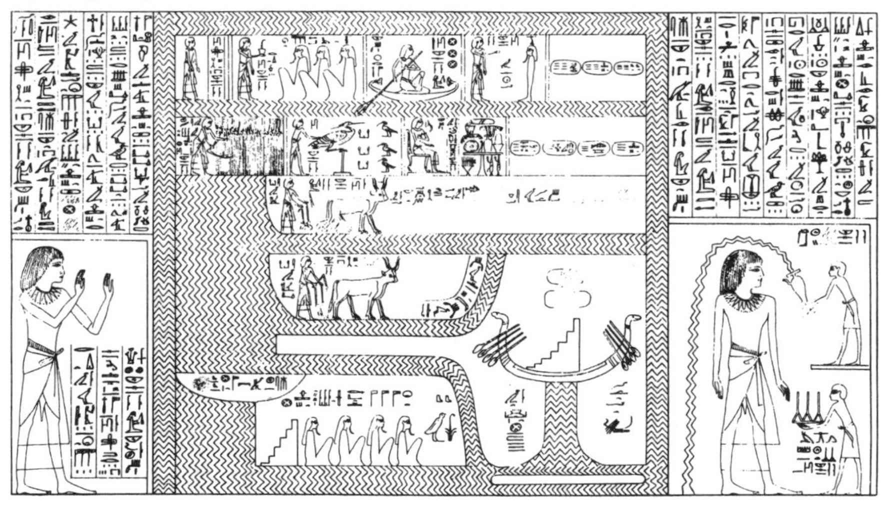

In the upper register we see three ovals named "Battle (field)", "Offering (place)" and "Great (place)". Then we see Nebseny in front of a god (Amset ?). In the middle he is rowing to "places" (⊗ ⊗ ⊗). In the next scene he is burning incense to gods, who represent the great Ennead. In the left corner we see just Nebseny. In the papyrus of Yuya the deceased is reaping here, too; obviously a harmonization in view of the two ploughing scenes. The four ovals under the three just mentioned are called "Offering (place)", "Red (place)", "Green (place)" and "Mistress of Both Lands". Further down we see a ship with a staircase on a "mound of water".

25 Upper ploughing scene (damaged line).

Generally it is called "sustenance-ship" (*ḏf3t*); here we have a "snake-ship" (*ḏdft*).[26] Another staircase can be seen behind four gods who form "the great Ennead that is in the Offering Field". In the former Coffin Texts these gods were identified as the four Sons of Horus, who especially protect the viscera. Therefore they secure from hunger and thirst. In the segment above the staircase is written "Birthplace of god" and "Battle (field)".

LEISURE OR LABOUR?

Our understanding of ancient religions may inadvertently be troubled by a Western way of thinking and a Christian way of believing. We are inclined to consider the Book of the Dead to be a kind of traveller's guide for the hereafter, aiming at the deceased's safe arrival in a paradise, where he may enter upon the everlasting joys of heaven. But the eschatological conceptions of the great monotheistic religions are not entirely compatible with ancient Egyptian ideas concerning the hereafter. Eating, drinking, making love may indeed be enjoyed. But what about labouring fields? Would that really contribute to eternal bliss? The Egyptians apparently had their doubts and made precautions. They secured the help of *shabtys*, little statues serving as substitutes for the deceased in case of conscription of labour in the hereafter.

That, however, does not alter the fact that the field-labourer in the vignette is none other than the beneficiary of the papyrus. "Reaping in the Offering Field by the scribe Nebseny" is written behind the man with the sickle. And the ploughing man can be identified likewise. In other manuscripts, too, it is the beneficiary of the papyrus who is doing the work; men as well as women. Even king Ramesses III and queen Nozemt did not scruple to have themselves depicted doing this humble, heavy labour.[27] The only exception seems to be king Pinozem I. In his Book of the Dead he had himself replaced by an anonymous servant; deliberately, it would seem.[28]

There is another conspicuous thing about these labourers. Their outfit is not quite appropriate for the kind of work they are supposed to do. King Ramesses III, for instance, is clad in a ceremonial dress including the white crown while ploughing and reaping. The fact that the deceased is clean and well-dressed is emphatically shown on the right side of the Offering Field. Nebseny is purified by a servant, while another presents him clothing. Hornung distinguishes between ploughing and reaping on the one hand and inferior work like irrigating and dunging on the other. The inferior part of the job was to be done by *shabtys*.[29]

---

26 The sungod travels in a snake-ship through the realm of "Sokar, who is on his sand". See E. Hornung, *Das Amduat. Die Schrift des verborgenen Raumes*, 3 vols. (Wiesbaden, 1963-1967), I: Vierte Stunde; Fünfte Stunde.

27 Heerma van Voss, "Ploegen", 114, note 2.

28 M. Heerma van Voss, "Religion und Philosophie im Totenbuch des Pinodjem I", in *Religion und Philosophie im alten Ägypten*, ed. U. Verhoeven, E. Graefe, (Leuven, 1991), 155-157.

29 E. Hornung, *Das Totenbuch der Ägypter*, (Zürich, München, 1979), 483.

But why would a well-to-do owner of a Book of the Dead, who in all probability never degraded himself to any kind of field-work during his earthly life, suddenly take pleasure in ploughing and reaping? Scribes like Nebseny despised peasant's work, as we learn from a writer's encomium:

> By day he cuts his farming tools; by night he twists rope. Even his midday hour he spends on farm labor. He equips himself to go to the field as if he were a warrior. ...
> When he reaches his field he finds it broken up. He spends time cultivating, and the snake is after him. It finishes off the seed as it is cast to the ground.[30]

Probably the agricultural activities of Nebseny and others are just ceremonial, like the kick-off of a high-heeled lady at a charity match. The job itself is thought to be done by *shabtys*. This is quite imaginable, considering that temple-scenes in which the king figures as a priest, generally reflect an ideal situation, the actual ceremonies equally being performed by substitutes, the professional priests. Heerma van Voss's opinion that the deceased required *shabtys* for all duties therefore seems plausible; at the same time, the dogmatic tradition remained uninjured, because of the representation of the deceased himself in text and vignette.[31]

"HERE WE HAVE NO LASTING CITY"

If the Offering Field were the deceased's ultimate destination, BD 110 should always conclude the series of spells in copies of the Book of the Dead. This is, however, not the case, although some Ramesside and later manuscripts end up with this spell. BD 110 was connected with a series of spells concerning the Powers of Sacred Places. Since the Offering Field was located in the east, BD 110 was linked up with a "spell for knowing the Eastern Powers" (BD 109). But there are also Western Powers (BD 108) and Powers of ancient cities like Buto (BD 112), Hieraconpolis (BD 113), Hermopolis (BD 114) and Heliopolis (BD 115).

What is more, in the Book of the Dead, the Offering Field is connected with the Field of Rushes; an area located in the east, too. By then, both regions can hardly be distinguished any more. The *parallellismus membrorum* in the title of BD 110, cited above, may indicate that. In BD 149, a spell dealing with fourteen different netherworld regions ("mounds"), the Field of Rushes occupies the second place (BD 149b), an indication in itself that this Field was not conceived as the deceased's final destination. Nor is it a "paradise" according to our

30 Papyrus Lansing (BM 9994). Translated by M. Lichtheim in *Ancient Egyptian Literature*, 3 vols., (Berkeley, 1973-1980), II: 170.

31 Heerma van Voss, "Ploegen", 113.

conceptions, as we will see. In the papyrus of Nu,[32] the text of BD 149b runs as follows:

> I am rich in possessions in the Field of Rushes.
> O, this Field of Rushes!
> Its walls are the firmament.[33]
> Its barley is five cubits high,
> two cubits its ear,
> three cubits its stalk.
> Its emmer is seven cubits,
> three cubits its ear,
> four cubits its stalk.
> Spirits, nine cubits tall, each of them all,
> reap it in the presence of Re-Horakhty.
>
> I know the gate in the middle of the Field of Rushes
> from which Re goes forth in the East of the sky;
> the south of it is the lake of the *Khar*-geese
> the north of it is the pool of the *Ro*-geese;
> the place where Re sails by wind, by rowing.
> I am a rigger in the God's ship,
> I am an indefatigable rower in the bark of Re.
>
> I know those two sycamores of turquoise
> between which Re goes forth,
> which grow on the supports of Shu
> at that gate of the Lord of the East,
> from which Re goes forth.
>
> I know this Field of Rushes of Re.
> Its barley is five cubits high,
> two cubits its ear,
> three cubits its stalk.
> Its emmer is seven cubits,
> three cubits its ear,
> five (*sic*) cubits its stalk.
> Spirits, nine cubits tall,
> reap it in the presence of Eastern Powers.

---

32 Budge 367,11 - 369,6. See also K. Sethe u. Gen., "Die Sprüche für das Kennen der Seelen der heiligen Orte", *ZÄS* 59 (1924), 1-20.

33 The spelling here corresponds with *bi3* in BD 85 (det. ◦◦◦), where the meaning obviously is "firmament"; see Naville, *Todtenbuch* II: 194 and Budge 185,6.

We learn that the deceased is rich in the Field of Rushes. This, however, does not mean that he will stay forever in the midst of his possessions. In the midst of the riches, we discern a gate. The symmetrical structure of the spell reflects what it is all about. The attention is drawn by the gate or the two sycamores in the centre, from which Re will go forth. Re is sailing by wind and by rowing in the middle of the Field, barley and emmer growing on either side. Spirits reap it when the sungod passes. This reminds one of the three registers of the hours in the "book" *Amduat*. The second hour in particular, with the introductory remarks: "Resting in *Wernes* by the Majesty of this God, rowing *Iaru*[34] in the waters of Re." In the middle register the sungod is depicted in his bark; on both sides we see the inhabitants of this second hour, some of them provided with corn.[35] The deceased Nu claims to know the gate or the two sycamores. This might be expected from an inhabitant who lives there eternally. But we are not concerned here with the deceased's knowledge of his whereabouts. What matters is knowledge that enables him to use the gate in the following of the sungod Re. Whether it is by wind or by rowing that Re goes by, the deceased forms part of the crew, either manipulating the rigging or rowing indefatigably. Therefore, he benefits from the agricultural activities of the huge spirits, who reap on either side in the presence of the sungod. So the important thing is not the acquisition of an eternal residence permit in the Field of Rushes. The important thing is to pass the Field of Rushes in the company of the sungod.

The gate in question is the gate of the eastern horizon. In the vignettes of BD 149b, this Field is symbolized by the sign of a mountain: . In the middle the name of the sungod has been written: , Re-Horakhty, "Re-Horus–of–the–horizon". Substituting the sun-disc for the sungod's name, we read , the well-known hieroglyph for the eastern horizon. In other words, the Field of Rushes is the area of the eastern horizon. The walls of this area, which appeared to be as extensive as the sky, can be seen in the firmament.[36] Precisely here, on the border of world and netherworld, we have the crucial point in religious topography: the place of resurrection. The deceased longs for this area in order to participate in the daily resurrection of the sungod. In other words, he longs for it in order to pass it eventually.

CHARACTER OF THE HEREAFTER

In the description of BD 149b, the Field of Rushes is neither pleasant nor unpleasant. As to leisure, the deceased is better off here than in the Offering Field of BD 110, where he had to plough and reap himself. At the same time, however, BD 110 mentions some other activities which the deceased will certainly enjoy.

---

34 Referring to *sekhet iaru*, "Field of Rushes"? In the tombs of Useramun and Sethos II spelled , with the determination not only of a god, but also of a locality!

35 Hornung, *Amduat*, I: Zweite Stunde; II: 61.

36 See above, note 33.

Notwithstanding these assets, the region is not altogether paradisiac. The spell opens by mentioning the aggression of Seth against Horus and how the deceased gets involved: he identifies with *Hotep*, the god of the Offering Field. This allows him to play the part of peacemaker, for the word *hotep* not only refers to "offering", it also means "peace":

> He pacifies[37] both fighters for the sake of those who belong to life.[38]
> He creates what is good, he brings peace.[39]

A few lines further the deceased himself is speaking again, boasting of his magic. But it is like whistling in the dark. Magic was needed. A carefree life was not guaranteed, it would seem. Especially the aboriginal spirits were feared:

> My utterance is powerful, I am sharper than the spirits.
> They shall not have power over me.[40]

The projection of human life in the hereafter includes besides the pleasures of life also the grim parts: the threat of torment and execution:

> I provide for this Field of yours, O *Hotep*,
> which you love, which you made,
> the "Mistress of Winds".
> I am glorious in it,
> I eat in it, I drink in it
> I plough in it, I reap in it,
> I do <not>[41] perish in it, I make love in it.
> My (magic) words[42] are powerful in it.
> I am not aroused in it, I am not beaten in it.
> <My heart is happy in it,
> for I know>[43] the (execution-)post of *Hotep*:
> *Bequtet* is its name. ...[44]

Knowledge is power. The deceased cannot be submitted to torturous interrogation and execution in the Offering Field, because he knows the name of that jackal-headed post, , that suspects were tied up to: *Bequtet*. So the happiness of the heart may be understood as relief.

---

37 *śḥtp*, a causative verb derived from the root *hotep*.
38 Error or euphemism for *imnt*, "the west", *i.e.* the realm of the dead (cf. *CT* V 340b).
39 Budge 224,15-16.
40 Budge 225,6-7.
41 Emendation based on *CT* V 345b.
42 "My magic" (*ḥk3w.i*) in *CT* V 345c.
43 Corrupt; translation based on *CT* V 346a-b.
44 Budge 225,7-13.

None of the other "mounds" in BD 149 is described as a paradise either. On the contrary, most of them are inhospitable. After the Field of Rushes (149b) we reach a "Mound of Spirits" (149c), "over which cannot be sailed; it holds spirits; its flame is a flash of fire." The mound of 149d ("Very High Twin Mountains") lodges a huge snake, called "Caster of knives". "It lives from decapitating the spirits and the dead in the realm of the dead." The mound of 149e, another "Mound of Spirits", cannot be passed. "The spirits within have buttocks of seven cubits, they live on the shadows of the weary ones" (*i.e.* the dead). The sixth mound (149f) seems to be some watery abyss, called *Imhet*. The god within is called "He who fells the *ʿḏ*-fish". Apparently he is a threat for the deceased, too, judging from the conjuration "That one 'who fells the *ʿḏ*-fish' shall not have power over me; the interrogators shall not pursue me, the adversaries shall not pursue me". The next mound (149g) is probably a kind of reed-marsh, *Ises*, "too far to survey; its blaze is fire." There is a snake, called *Rerek*. "It is seven cubits long over its back. It lives on spirits, destroying their power." BD 149h deals with the mound *Ha-hotept*. It seems to be a "flood, over the water of which no one has power, because it is greatly feared because of its roaring". The vignette of 149i is significant of the inhospitability of this ninth mound.

Fig. 6 Ikesy

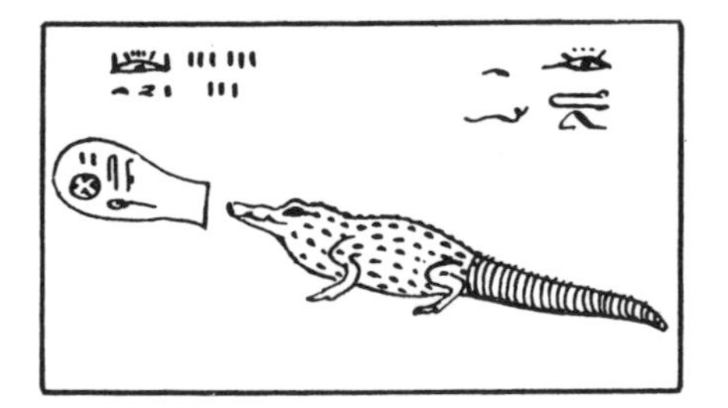

We see (Fig. 6) a crocodile lying in wait at the opening of something like a toppled jar. Above the animal we read: "watching what he is about to catch." The "jar" is characterized as the place *Ikesy*. The vignette unambiguously illustrates what is stated about this mound in the spell: "*Ikesy* that no one will leave who entered it", except the august god who is in his egg, whom gods fear and spirits dread; that is to say, the sungod. "Its opening is fire, its air is destruction to noses." As to the tenth mound with the name "Which is at the entrance of the district" (149k), the deceased fears that his power will be taken away and his shadow violated. *Idu* (149l) is a prison like *Ikesy*, "that no one will leave who entered it". It hides corpses and violates spirits. The "Mound of *Wenet*" (149m) is as fiery as *Ises*: "its blaze is fire". The following "Mound of Water" (149n) is hot as well, for "its water is fire, its waves are fire, its blaze is a flash of fire, lest its water will be drunk to quench their thirst that they will have, because it is greatly feared and highly awe-inspiring". The last spell in the series concerns the "Mound of *Kheraha*" (149o). Here, as well as in BD 149a ("Mound of the West") and 149b (concerning the Field of Rushes), grimness seems absent. Hospitality, however, is another thing. For the deceased has to beg or to command.

The mounds of BD 149 are obviously not inviting for settlement. On the contrary, several times we learn that the deceased is anxious to pass them.[45] BD 149b concerning the Field of Rushes belongs to the series.

TWO WAYS

This is in conformity with the place BD 149 generally occupies in the Book of the Dead. The concatenation of spells, generally followed by BD 150 (another series of vignettes of the mounds preceded by four snakes[46]), often concludes a manuscript. And especially the connection with the generally preceding BD 136B appears to be strong, as can be seen from the following list of manuscripts[47]:

BD 149 in manuscripts of the eighteenth dynasty (*c.* 1550-1300 BC):

pCairo of Yuya	... 136B – 149 – 150
pCairo of Amunhotep [Cc]	/// 149 – 150 – 17 – 18 /// 124 – 75 – 105
pCairo of Amunhotep [Cd]	... 136B – 149 ///
pParma of Amunhotep [Ig]	... 136B – 149
pCracow of Uia [Pp]	/// 149 – 150 – 42 ///
pCairo of Maiherperi	... 136A + 136B – 149 – 150 – 125 –
pLouvre of Mesemneter [Ca]	... 136B – 149 – 150 – 103 – 117 – 133
pBM 10477 of Nu	... 136B – 149 – 150
pBM 9900 of Nebseny [Aa]	... 136B – 149 – 150 – 125 – offering scene – 172
pCairo of Nebseny	/// 136B – 149 – 150 – 100 /// offering scene (*bis*)
pLouvre of Nebqed [Pe]	... 110 /// 149 – 150 – 79
pLouvre of Neferubenef [Pb]	... 136B – 149 – 150
pCracow of Neferhotep	/// 149 – 150 – 125 ///
pVienna of Sesostris	... 136B – 102 – 149 – 153A – 150 – 123 –
pLouvre of Thenna [Pc]	... 136B – 149 – 150
pCairo of Senhotep	... 136B – 147 – 125 – 149 – 150 – A.O.
pBM 9905 [Ac]	... 125 – 149 – 150
pBM 9913 + pTuy	... 136B – 149 – 150 – 125 –
pBM 9943	... 136B – 149 – 150 – 125 – 136B –
pBM 10009 + 9962,1 [Ad]	... 136A + B – 149 – 150 – 15/16 /// 147 ///

45 The same anxiety underlies the occurrence of the *senet*-game in the vignettes of BD 17: passing through the netherworld. See my "It is All in the Game" in *Funerary Symbols and Religion*, ed. J.H. Kamstra, H. Milde, K. Wagtendonk, (Kampen, 1988), 89-95.

46 See the sequence in the Theban tomb [353] of Senenmut, published by Hornung, *Totenbuch,* 317.

47 Based on the catalogue in I. Munro, *Untersuchungen zu den Totenbuch-Papyri der 18. Dynastie. Kriterien ihrer Datierung*, (London, New York, 1988), 274-308.
pCairo = papyrus Cairo. A.O. = Adoration of Osiris. Between [ ] the siglum in Naville, *Todtenbuch.*
/// damages; – connection of spells; ... omitted spells.

pBM 10489+Amsterdam+Boston+New York+Redwood Library+Stockholm
... 136B – 149 – 150 – ? – 126 – 125 –
pLouvre N.3097 [Pf] ... 136B – 83 – 30B – 85 – 77 – 149 – 150
pLouvre E.11085 ... 136B – 149 (?) ///

BD 149 in Ramesside manuscripts (*c.* 1300-1080 B.C.):

pBrussels of Neferrenpet	... 136B – 149 – 150 – 64 – 30B –
pBM 10471 / 73 of Nakht	... 42 – 149 – 64 – 150 – 151 – Hymn to Re
pBerlin 3002 of Nakhtamun	... 137A-B – 63A – 149 – 130 – 30B –
pLouvre (+BM) of Nozemt	... 136B – 149 – 150 – 125
pLeiden T.5 of Ray	... 175 – 149 – 186
pParis BN of Sutymes [Pd]	... 125 – 110 – 149 – 16B
pMilano (pBusca)	... 110 – 144 – 146 – 149 – 29B(?) – 30B –
pLeiden T.6 [Lc]	... 110 – 149 – 125 ///
Coll. Mallet [Ph]	... 110 – 149 – 77 – 81 –

One may say that BD 136B is a kind of introduction to the spells of the different mounds in BD 149.[48] It is a "spell for sailing in the great bark of Re and for passing the circle of fire". That reminds us of the many fiery regions of BD 149, which follow. Originally "the circle of fire" must be understood as a court surrounding Re, for BD 136B is composed of two Middle Kingdom Coffin Texts, the spells 1033 and 1034. *CT* 1033 is a "spell for passing the entourage of fire of the cabin of the bark of Re".[49] Some coffins have the word *šnwt*, "circle" determined with or , suggesting the meaning "entourage", "court". According to Hornung, *CT* 1034 is "ein Schlangenzauber, der vor den Uräen des «feurigen Hofstaats» schützen soll, die den Sonnengott als feurige Aura umgeben".[50] *CT* 1033 is written within a rectangle. *CT* 1034 follows outside. On the coffin B1C the four red bands of the rectangle itself are inscribed with , "entourage of fire".[51] It represents "den 'feurigen Hofstaat' des Sonnengottes mit einem gleichzeitig schützenden und abwehrenden Charakter".[52] Later it may have been understood as a reference to the fiery regions of BD 149 with their severe watchmen, especially since *CT* 1034 ends up with the title "Guidance to the paths of *Rostau*".

In the *CT* these paths are represented in a sort of map. We see a blue waterway and a black path overland. The two ways are separated by a "Lake of Fire". All along there are regions watched by demons. One of the regions is the Offering Field (*CT* 1048-1052). It looks like the ancient editors of the Book of the Dead substituted BD 149-150 for the regions along these paths of *Rostau*. The strange

48 See also H. Milde, *The vignettes in the Book of the Dead of Neferrenpet*, (Leiden, 1991), 106-130.
49 *CT* VII 278a-b.
50 Hornung, *Totenbuch,* 498.
51 *CT* Sp. 1032.
52 E. Hermsen, *Die Zwei Wege des Jenseits*, (Göttingen, 1991), 236; see also 108-109.

redoubling of the vignettes of BD 149 in BD 150 may be explained as an adaptation of the two ways in the new setting. These vignettes should not be seen one after the other, but next to each other, like the parallelism of the two ways. And the unaccountable occurrence of the four snakes at the beginning of BD 150 may be understood as well. For it should be noted that *CT* 1034 is a "Schlangenzauber". It recurs at the end of BD 136B:

> On your faces, you snakes yonder, let me pass!
> I am strong, lord of strong men.
> I am a noble of the lord of righteousness,
> whom the cobra-goddess made.
> My protection is the protection of Re
> and look, he went round in the Field of Offerings,
> a God greater than you,
> who counts his Ennead among those who give offerings.[53]

The snakes of BD 150 possibly mark the transition from the end of BD 136B to the double series of mounds.

In short, BD 149-150 is the New Kingdom variation of the Book of the Two Ways. The roads through the netherworld regions, both gloomy and paradisiac, lead to the resurrection point in the eastern horizon. The deceased, once introduced into the realm of the dead under the rule of Osiris, has entered a cosmic circuit in which he travels with the sungod Re, partaking in his daily resurrection.

Fig. 7 Ceiling in Sennezem's tomb-chamber

Finally, it is good to note that time is something different in the hereafter. We learn from "books" like *Amduat* that every hour means a lifetime for the inhabit-

53 Budge 302,15 - 303,7.

ants.[54] So the deceased's sojourn in the Fields of Rushes is not necessarily short. He may enjoy his riches extensively. However, the ultimate concern remains the participation in the sungod's resurrection, no matter how paradisiac the Field of Rushes or the Offering Field may be. In his tomb-chamber (Fig. 7) we see Sennezem and his wife not only depicted in the luxurious east, but also in the west and in between, along the ceiling; that is to say: along the sky, adoring the gods of heaven and the Powers of East and West, of Sacred Places like Hermopolis and Hieraconpolis, opening the gate of heaven, witnessing the triumph of Re, while they are nourished by the sky-goddess Nut vaulted over them as a tree (Fig. 7).

ABBREVIATIONS

Budge E.A. Wallis Budge, *The Book of the Dead. The Chapters of Coming Forth by Day. The Egyptian Text in Hieroglyphic Edited from Numerous Papyri*, London, 1898.

CT A. de Buck, A.H. Gardiner, *The Egyptian Coffin Texts*, 7 vols., Chicago, 1935-1961.

JEA *Journal of Egyptian Archaeology.*

JEOL *Jaarbericht van het Vooraziatisch-Egyptisch Genootschap Ex Oriente Lux.*

LÄ *Lexikon der Ägyptologie*, ed. W. Helck, E. Otto†, W. Westendorf, 7 vols., Wiesbaden, 1975-1992.

MDAIK *Mitteilungen des Deutschen Archäologischen Instituts, Abteilung Kairo.*

RÄRG H. Bonnet, *Reallexikon der ägyptischen Religionsgeschichte*, Berlin 1952.

ZÄS *Zeitschrift für Ägyptische Sprache und Altertumskunde.*

54 E. Hornung, *Geist der Pharaonenzeit*, (Zürich, München, 1989). 73; E. Hornung, *Die Nachtfahrt der Sonne*, (Zürich, München, 1991), 79-80.

BIBLIOGRAPHY

J. Assmann, "Tod und Initiation im altägyptischen Totenglauben" in *Sehnsucht nach dem Ursprung*, ed. H.P. Duerr, (Frankfurt, 1983), 336-359.
P. Barthelmess, *Der Übergang ins Jenseits in den thebanischen Beamtengräbern der Ramessidenzeit*, Heidelberg, 1992.
G. Goyon, *Rituels funéraires de l'ancienne Égypte*, Paris, 1972.
M. Heerma van Voss, "Religion und Philosophie im Totenbuch des Pinodjem I", in *Religion und Philosophie im alten Ägypten*, ed. U. Verhoeven, E. Graefe, (Leuven, 1991), 155-157.
M. Heerma van Voss, "Van ploegen tot aren lezen", in *Beginnen bij de letter Beth*, ed. K.A. Deurloo, F.J. Hoogewoud, (Kampen, 1985), 111-114.
E. Hermsen, *Die Zwei Wege des Jenseits*, Göttingen, 1991.
E. Hornung, *Ägyptische Unterweltsbücher*, Zürich, München, 1972.
E. Hornung, *Altägyptische Höllenvorstellungen*, Berlin, 1968.
E. Hornung, *Die Nachtfahrt der Sonne,* Zürich, München, 1991.
E. Hornung, *Das Totenbuch der Ägypter,* Zürich, München, 1979.
Chr. Jacq, *Le voyage dans l'autre monde selon l'Égypte Ancienne*, Paris, 1986.
H. Kees, *Totenglauben und Jenseitsvorstellungen der alten Ägypter*, Leipzig, 1926; 3rd ed., Berlin, 1977.
L.H. Lesko, *The Ancient Egyptian Book of the Two Ways*, Berkeley, Los Angeles, London, 1972.
E. Lüddeckens, *Untersuchungen über religiösen Gehalt, Sprache und Form der ägyptischen Totenklagen*, Berlin, 1943.
H. Milde, "It is All in the Game" in *Funerary Symbols and Religion*, ed. J.H. Kamstra, H. Milde, K. Wagtendonk, (Kampen, 1988), 89-95.
S. Morenz, "Der Tod und die Toten", Chapter IX in *Ägyptische Religion*, (Stuttgart, 1960), 192-223.
S. Morenz, "Ägyptischer Totenglaube im Rahmen der Struktur ägyptischer Religion", *Eranos Jahrbuch* 34 (1965), 399-446.
E. Otto, *Das ägyptische Mundöffnungsritual*, Wiesbaden, 1960.
A. Piankoff, *The Wandering of the Soul*, Princeton, 1974.
C.E. Sander-Hansen, *Der Begriff des Todes bei den Ägyptern*, Kopenhagen, 1942.
Chr. Seeber, *Untersuchungen zur Darstellung des Totengerichts im Alten Ägypten*, München, 1976.
J. Settgast, *Untersuchungen zu altägyptischen Bestattungsdarstellungen*, Glückstadt, Hamburg, New York, 1963.
A.J. Spencer, *Death in Ancient Egypt*, Harmondsworth, 1982.
J. Spiegel, *Die Idee vom Totengericht in der ägyptischen Religion*, Glückstadt, Hamburg, New York, 1935.
J. Yoyotte, "Le jugement des morts dans l'Égypte ancienne" in *Le jugement des morts*, Sources Orientales IV, (Paris, 1961), 15-80.
J. Zandee, *Death as an Enemy according to Ancient Egyptian Conceptions,* Leiden, 1960.

Death as a Privilege
The Hittite Royal Funerary Ritual*

Theo P.J. van den Hout

INTRODUCTORY REMARKS

The institution that we call "kingship" was – and still is – in many cultures and societies a binding factor of importance, representing the unity of a nation and the personification of its collective power. Tradition is an important aspect in royal ideology: a ruling house likes to present its foundations as deeply rooted in national history, looking back on a long and preferably uninterrupted line of predecessors. Taking on the leadership over a country, however, implies responsibility, so that often the fate of the leader is bound up with the well-being of 'his' country. This may not so much mean that the king is easily criticized when something goes wrong but rather that he is the embodiment of the "prosperity and perpetuity of the political order."[1] As long as he is strong and healthy, the country is, too. In its modern form we can see this principle at work when the Dow Jones index and the exchange rates of a nation's currency keep pace with the health and well-being of the world's leaders. As a consequence, the death of a king may not only create a politically unstable situation, but can be highly disconcerting from an ideological point of view as well. A successor, therefore, is often quickly enthroned to discourage any potential usurpers, and in the funeral ceremony of his predecessor the idea of continuity and perpetuity of kingship is stressed. The deceased king's body is embalmed or a lifelike effigy is made to deny, as it were, his death. Theories such as that of "the king's two bodies" are developed: his "body natural" and his "body politic."[2] The former refers to the individual king as a human being subject to all possible physical defects, and the latter to him as the abstract office of government carried on by one individual after another: the king as guardian of the Crown. So, to safeguard the country at the moment of a king's demise, emphasis has to be laid on visualizing that body politic. But also in a situation where there is no threat of interregnal disorder whatsoever, the death of a

---

* Research for this study was made possible by a fellowship from the Royal Dutch Academy of Arts and Sciences. Furthermore, I would like to express my gratitude to Ph.H.J. Houwink ten Cate for his most valuable suggestions. To Dr.-Ing. P. Neve I am indebted for his kind permission to publish here the two photographs of the miniature axes (Figs. 5-6) as well as the map of Ḫattuša (Fig. 1) and to Dr. D.J.W. Meijer for the photographs of Yazilikaya Room B.

1 R. Huntington – P. Metcalf, *Celebrations of Death. The Anthropology of Mortuary Ritual* (Cambridge 1979), 154.

2 R. Huntington – P. Metcalf, ibid. 159-165.

ruler is an opportunity to reinforce royal ideology: "It seems that the most powerful natural symbol for the continuity of any community, large or small, simple or complex, is, by a strange and dynamic paradox, to be found in the death of its leader, and in the representation of that striking event."[3]

All this is true for Hittite society in Anatolia during the second millennium BC as well. Their kings used to introduce themselves in their official texts with a genealogy of sometimes over four generations, occasionally claiming descent from the almost legendary founders of Hittite rule. They practised a royal ancestor cult with regular offerings to kings and queens reaching back three centuries or more. Having got the land from the gods to govern, the king was considered responsible for its well-being. This made him an intermediary between god and man, representative of the Sungod on earth. Although close to the gods, he was not considered divine himself, but turned into a god at death.[4] Queens, princes and princesses allegedly shared this fate.[5] However, apart from the common expression of "becoming a god" at a king's death, there are only few, but often quoted, passages that explicitly refer to his divine fate. Instead, there is strong evidence that the Netherworld, too, played an important role in a king's death.

The restriction of the theme for this contribution to ideas about death and immortality in *court* circles was not a deliberate and free choice. Our Hittite sources come exclusively from temple and palace archives and hence inform us predominantly about life in the upper echelons of that society. Moreover, it seems that most of the key positions, whether military, administrative or religious, were held by a network consisting of members of the royal family including nephews, cousins and in-laws. Remarks concerning the death and possible beliefs relating to an afterlife of officials may therefore often be likely to reflect royal ideology as well. As a result, we hardly know anything about the role death and immortality played in the lives of the 'ordinary' citizens of the Hittite empire. Here archaeological material is a most valuable additional source.

One of our main sources for beliefs held by the former group is the description of an intricate royal funerary ritual lasting for two weeks and heavily laden with symbolism. This *rite de passage* has been preserved fairly well among the tens of thousands of cuneiform tablets excavated at the site of the ancient Hittite capital of Ḫattuša, nowadays the small Turkish village of Boğazkale (Boğazköy). Most of the fragments constituting this composition have been known for quite some time, and a first edition of these texts appeared almost 40

---

3 R. Huntington – P. Metcalf, ibid. 182.

4 During the 13th century BC, however, a tendency seems to have grown to deify the Hittite king already during his lifetime or to confer upon him at least certain privileges that were formerly only reserved for gods, cf. P. Neve, *Ḫattuša – Stadt der Götter und Tempel. Neue Ausgrabungen in der Hauptstadt der Hethiter* (Mainz 1992), 6, 85, and Th.P.J. van den Hout, *Tudḫalija Kosmokrator. Gedachten over ikonografie en ideologie van een Hettitische koning* (Amsterdam 1993), 20, 30-32.

5 See below §6. Although in the following pages we will mostly talk about the king, it must be kept in mind that his fate may have been largely representative of the royal family as a whole.

years ago from the hand of Heinrich Otten. One passage especially aroused the interest of a wider public because of the apparent parallels with the funeral scenes of Patroclos and Hector in the Iliad. Since Otten's edition of 1958 and two addenda shortly afterwards[6], some reviews[7], a French translation[8] and short overall treatments of the ritual or references to it from a religious-historical point of view[9] were published. The recent publication, however, of some additional fragments and the recognition of some others as belonging to this text again contribute to a better understanding of the ritual.

In the following we will review some general ideas about death and immortality as they emanate from the texts and contrast them with the archaeological record. Finally, we will examine more closely the royal funerary ritual just mentioned; a translation of the description of the events on the eighth day will serve as an example.

THE INEVITABILITY OF DEATH AND SOME OF ITS DESIGNATIONS

The notion of a predestined life span was well known among the Hittites.[10] In the Old Hittite ritual for the building of a palace that concerns as much the 'building' of the king we encounter the Hittite Parcae, the so-called *Gulšeš*-deities. The verb *gulš-* is used for inscribing, engraving or cutting: thus, they are the deities who already at birth have laid down the number of years of a king's life. Like their classical counterparts they spin the thread representing that life: "(One) has a distaff, they have full spindles. They are spinning the years of the king, and the years'end (and) their number cannot be seen."[11] If a person dies young, the *Gulšeš* may be blamed accordingly. A thirteenth century prince named Tattamaru was married to the niece of the queen. Apparently, their marriage was of only short duration because of her dying at a relatively young age. So the queen in a letter to the widower expressed herself thus: "You, Tattamaru, had married the

6 "Eine Lieferungsliste zum Totenritual der hethitischen Könige," *Welt des Orients* 2 (1954 1959), 477-479, and "Zu den hethitischen Totenrituale," *Orientalistische Literaturzeitung* 57 (1962), 229-233.

7 A. Goetze, *American Journal of Archaeology* 64 (1960), 377-378, E. Laroche, *Bibliotheca Orientalis* 18 (1961), 82-84, R. Werner, *Orientalia* 34 (1965), 379-381.

8 L. Christmann-Franck, "Le Rituel des Funérailles Royales Hittites," *Revue Hittite et Asianique* 29 (1971), 61-111.

9 Cf. J. Borchardt, "Epichorische, gräko-persisch beeinflußte Reliefs in Kilikien" *Istanbuler Mitteilungen* 18 (1968), 179-186, O.R. Gurney, *Some Aspects of Hittite Religion* (Oxford 1977), 59-63, id., *The Hittites* (Harmondsworth 1990), 137-140, E. Masson, *Les Douze Dieux de l'Immortalité* (Paris 1989), passim.

10 Cf. E. von Schuler in *Wörterbuch der Mythologie*, ed. H.W. Haussig (Stuttgart 1965), 168-169, 192-193, H. Otten – J. Siegelová, "Die hethitischen Gulš-Gottheiten und die Erschaffung der Menschen," *Archiv für Orientforschung* 23 (1970), 32-38, and H.A. Hoffner, "Hittite Terms for the Life Span," in *Love and Death in the Ancient Near East. Essays in Honor of Marvin Pope*, ed. J.H. Marks – R.M. Good (Gilford, Connecticut 1987), 53-55.

11 KUB XXIX 1+ ii 6-10 (CTH 414, ed. G. Kellerman, *Recherche sur les Rituels de Fondation Hittites* (diss., Paris 1980), 13, 27).

daughter of my sister, (but) then the *Gulšeš* treated you badly, and she died on you."[12]

Mostly, however, such an untimely death was not blamed on the gods but felt as a reaction on the gods' part to human behavior in the past. If the cause was not immediately obvious, it had to be established by way of oracles. In the case of a king, the whole country might suffer from divine wrath: in accordance with the Hittite ideology of kingship, the well-being of the state was bound up with the well-being of the ruler. As a consequence, any failure on his part could ultimately even result in a massive loss of lives among the population. Added to the concept that sins can be inherited, some wrongdoing of times long since passed could cause death in the present and times to come. This is exemplified in the well-known story of the epidemic that for over two decades oppressed central Anatolia during the last years of Šuppiluliuma I into the reign of his son and second successor Muršili II. According to the Hittite sources themselves this plague was brought into the country by "Egyptian" captives, but the cause of the divine wrath, which had taken on the form of the epidemic, was after research into their archives reduced to the following three events: an inner dynastic murder during and involving Šuppiluliuma I, the neglect of the cult devoted to the river Mala (i.e. the Euphrates) for many years and, finally, the violation of a treaty the Hittites had formerly concluded with the Egyptians.

But however inevitable, the fact that the time of death was unknown to mortals gave rise to the very human hope that, to a certain extent, death could be manipulated. Originally, for instance, Ḫattušili III (1267-ca. 1240) was said to be predicted to have only a short life. In a dream the goddess Ištar told his father to make his son her priest, which would secure him a longer life. Ḫattušili's wife, the queen Puduḫepa, not sure on the one hand when death would attack but convinced that it might be influenced on the other, made many a vow for her husband's well-being and longevity. In one of these, she addressed herself to the Queen of the Netherworld, the goddess Lelwani:

> If you, Lelwani my Lady, will intercede with the gods, (if) you will keep your servant Ḫattušili alive (and) grant him long years, months (and) days, then I will order to be made for Lelwani, my Lady, a silver statue of Ḫattušili, as tall as Ḫattušili, (with) his head, his hands (and) his feet in gold.[13]

One notable exception to the inevitability of death involving the king can be seen in the so-called substitution rituals. Although these rituals have to be traced back to Mesopotamian origins and their borrowing seems to have taken place at a relatively late date in Hittite history, i.e. not before the 14th century, these texts did

---

12 KUB XXIII 85, 5-6 (CTH 180, cf. A. Hagenbuchner, *Die Korrespondenz der Hethiter. 2. Teil. Die Briefe mit Transkription, Übersetzung und Kommentar* (Heidelberg 1989), 15-16).

13 XXI 27+546/u iii 36-41 (CTH 384, cf. D. Sürenhagen, "Zwei Gebete Ḫattušilis und der Puduḫepa," *Altorientalische Forschungen* 8 (1981), 116-117).

not merely belong to the scribal curriculum but were really put into practice.[14] According to the colophon of one of them,[15] the ritual was performed in case a king's death was presaged through a dream, extispicy, auspicy or some evil omen. The best preserved of the substitution rituals that have come down to us was triggered off by an unspecified lunar omen. Some of the mantic texts give us an idea of what sort of omen it may have been. In the traditional formulary of protasis ("If ... ") and apodosis ("then ... ") we find predictions like the following: "If in the third month on the fourteenth day the Moon dies (i.e. eclipses), then the king will die, his son will seize [the throne], floods will rise, ... or the troops will fall in masses."[16] Another reads: "[I]f on the twentieth day (i.e. of the seventh month[17]) the Moon dies: the king<'s> days are cut short, destruction of the country"[18], and further on in the same text: "If on the twentieth day (i.e. of the eighth month[19]) the Moon dies, then the king will die in battle."[20]

Once such an omen had appeared and its fateful meaning had been established by the mantic specialists at the royal court, preparations for a ritual were begun. In the substitution ritual just mentioned a prisoner was taken and, at nightfall, anointed with the oil of kingship while the king said:

> See, this man here is king. [I have bestowed] on this man the name of kingship and I have dressed this man with [the garment of ki]ngship as well, and I have put on him the *lupanni*.[21] Evil omen, short years, short days take note [of this man] and go after this substitute![22]

Some official of the Hittite king then escorted the prisoner-king back to his country. At dawn the next morning, the 'old' king went through purification rites and prayed to the Sungod and Sungoddess in heaven and the goddess of the Netherworld Lelwani to accept the substitute king.

In spite of this possible way-out, one of the names for death was "the day of one's destiny". This euphemism seems to have been an ordinary way of referring to someone's future death. It is attested and known to have been used in treaties in stipulations concerning succession rights: "When for you the day of your destiny will have come, (they will take your son and install him as king)."[23]

---

14 H.M. Kümmel, *Ersatzrituale für den hethitischen König* (Wiesbaden 1967; = *StBoT* 3), 188-191.
15 KUB XV 2 rev. 5'-9' (w. duplicates; CTH 421, ed. H.M. Kümmel, ibid. 70-71).
16 KBo VIII 47 obv. 9'-11' (CTH 532, cf. K.K. Riemschneider, *Die hethitischen und akkadischen Omentexte aus Boğazköy* (unpublished manuscript), 26-27; the fragments 76/g and 833/f have now been published as KBo XXXIV 110 and 111).
17 Cf. ibidem ii 14'.
18 KUB VIII 1+ iii 1-2 (CTH 532, ed., K.K. Riemschneider, *Omentexte* 100-107).
19 Cf. ibidem iii 8.
20 Ibidem iii 15.
21 I.e. a small, tightly fitting cap worn on the head by kings.
22 KUB XXIV 15+KUB IX 13 obv. 20'-24' (CTH 419, ed. H.M. Kümmel, *StBoT* 3, 10-11).
23 KBo IV 10+ obv. 5' (cf. Th.P.J. van den Hout, *Der Ulmitešub-Vertrag. Eine prosopographische Untersuchung* (Wiesbaden 1994; = *StBoT* 38), 20-21).

Even more concealing is the term "the good day". Although it never seems to be used in the direct sense of the "day on which one dies/will die", it is one of the more important expressions used in the ritual for a deceased king (see below) and usually taken to designate the moment of death. Sometimes it is accompanied by the divine determinative marking it as a deity able to receive offerings, just like the Sungod of Heaven, the Sungoddess of the Earth and the Soul of the Deceased.

Significant in the sense that it may tell us something about the way in which the Hittites conceived of the process of dying, is a third expression for death: "the day of (one's) mother." Apart from some attestations in treaties and oracles[24], in which it seems to be a mere equivalent of, for instance, "the day of one's destiny", the following passage from a funerary ritual is more revealing:

> [Th]en, the *patili*-priest, who (is) on top of the roof, calls down into the house and repeatedly calls the one who died by his name: 'Wher[e] has he gone?' Those gods with whom he (is), each say from bel[ow up (to the roof)]: 'To the *šinapši*-house he went' and the first one [from] the r[oof] calls down: 'Where has he gone?' Those gods with whom he then (is), sa[y] from below up (to the roof): 'Thereto [he w]ent.' When they finish the round of gods, then the clothes, that (are?) his [ ] (they?) say from below up (to the roof): 'Here he went or [ ].' He speaks from the roof downwards six times, [they] spe[ak] up(wards) six times. When for the seventh time he says downwards: 'Where has he g[on]e?', they say to him from below up (to the roof): 'For him the day of (his) mother [has come and] she has taken him by the hand and accompanied him.' [25]

In this scene the man on top of the roof probably represents the present world, those below the deities of the Netherworld.[26] Such scenes may have been acted out as part of the funeral rites or games. An unfortunately rather fragmentary but related text[27] shows that the king and queen were – or at least could be – the

---

24 Cf. KUB XXI 1+ i 64 (CTH 76, cf. J. Friedrich, *Staatsverträge des Hatti-Reiches in hethitischer Sprache*, 2. Teil (Leipzig 1926), 54-55) "[When for you] Alakšandu will have come the day of your mother"; KUB V 3 i 45-46 "If out of 10 (or) 20 people for some the day of the mother [will have come], but no massive dying takes place in Ḫattuša."

25 KUB XXX 28 rev. 1-12//KBo XXXIV 80, 1'-8' (cf. H. Otten, *Hethitische Totenrituale* (Berlin 1958), 96-97); the new duplicate KBo XXXIV 80, 6' (-*z*]*i an*[-*n*]*a-aš-ša-aš-ši* UD-*az*[) enables us now to recognize this expression here, that was restored differently hitherto, cf. H. Otten, *HTR* 17, 96-97, J. Puhvel, "Hittite *annaš šiwaz*," *Zeitschrift für vergleichende Sprachforschung* (= *KZ*) 83 (1969), 59, G. Beckman, *Hittite Birth Rituals* (Wiesbaden 1983; = *StBoT* 29), 236, and A. Archi, "Divinités Sémitiques et Divinités de Substrat. Le Cas d'Išḫara et d'Ištar à Ebla," *Mari Annales de Recherches Interdisciplinaires* 7 (1993), 77-78.

26 Cf. J. Puhvel, *KZ* 83 (1969), 59; L.-V. Thomas, "Funeral Rites," in *Encyclopedia of Religion*, ed. M. Eliade [16 vols.] (New York-London 1987), 5: 453, mentions a similar rite in China, in which a close relative of the deceased "climbs the roof of the house to 'call back his soul'", in order to make sure he is really dead: "if it does not return, there is no doubt about his death."

27 Cf. H. Otten, KUB XXXIX Inhaltsübersicht v.

object of this conversation. Given the restrictions mentioned at the beginning of this article concerning the social range of the Hittite text material, this is hardly surprising. Not having been transliterated before, the fragment (KUB XXXIX 49) is worth quoting here:

x+20 *kat-t*]*a-an-da ḫal-za-i* LUGAL-*uš-wa k*[*u-ua-pí*
21 *kat-*]*ta-an ša-ra-a ḫal-za-i* [
22]x-*ki-zi ma-a-an* MUNUS-*an-za-ma nu* x[
23]*ŠA* MUNUS-*TI-wa-za ar-ḫa a-ra-ši-iš*[
24]x-*az kat-ta-an-da ḫal-za-i* LUGAL-*uš*[
25 *kat-ta-an ša-*]*ra-a* ^LÚ^AZU *ḫal-za-i ku-it-wa-ra-*x[
26 -]*ki-mi ŠA* AMA-*ŠU-wa-aš-ši* UD.KAM-*za*[
27 -]*da-aš nu-wa-ra-an-za-an ták-na-aš* ^d^UT[U-
28]*nu-wa-ra-an* ŠU.[ḪI.A^?^] *ḫar-z*[*i*

[... do]wn he (i.e. probably the colleague of the divination priest[28]) calls: 'W[here has$^?$] the king [... '
[... from be]low upwards he (i.e. the divination priest[29]) says:
['...] he will [... '$^?$], but if (it is) a woman, then [
[...] from the woman's [...] his colleague[
[...] from the [...] down he calls: 'The king [...
[... from below u]pwards the divination priest calls: 'What [
[...] I will. For him the day of his mother [has come ...
[...] she$^?$ has, and him (from/to$^?$) the Sungoddess of the Earth [...
...] and she holds him by the hand[s$^?$ "][30]

The underlying idea of this expression may be that the mother who set her child on this earth is the one who will accompany him to his otherworldly existence also.[31] Interesting in this respect is the appearance of the *patili*-priest in the first passage: he is predominantly attested in birth-rituals as a purificatory priest. In all likelihood, therefore, "death is understood here as a kind of birth", as Beckman suggests.[32] It is very unfortunate that the last lines of the above fragment are so damaged that we cannot be sure about the exact role of the Sungoddess of the Earth, queen of the Netherworld, here.[33] The similar scene involving the *patili*-priest on the roof and the deities below takes place at the end of the day; in the lines immediately following, the Sungoddess of the Earth is offered food and

28 Cf. below line 23'.
29 Cf. below line 25'.
30 KUB XXXIX 49, 20'-28' (CTH 450).
31 This possibility is already considered by G. Beckman, *StBoT* 29, 236.
32 G. Beckman, *StBoT* 29, 237, for his discussion of the *patili*-priest see ibid. 235-238. The close link between death and birth seems to be almost universal: see the remarks by L.-V. Thomas in *Encyclopedia of Religion*, 5: 452, 453.
33 Cf. E. von Schuler, *WbMyth.* 199-200; she may have been either subject of a verb meaning "to let go/release" vel sim. or the person from whom the deceased was taken away (^d^UT[U-*i* ?).

drinks, and then a new day starts. So, if the gods "with whom he then (is)" are indeed the deities of the Netherworld, then his mother must be coming to fetch him, i.e. the king, for his ascent to heaven, and the offerings to the Sungoddess of the Earth were probably meant to appease her and to set the king free. If so, this could indeed confirm that also in the first fragment it is a royal person who is fetched by the mother, because persons of royal descent seem to have been the only ones entitled to escape the Netherworld, that is, the mother apparently able to enter and leave as she pleases, and the king himself. This is a question we will have to return to.

CONTACTS WITH THE SOUL OF THE DEAD

At the point of death body and soul diverge: the body either decays or is cremated but the soul, after having entered the body at birth, leaves again. This 'soul', as it is traditionally rendered, is the means through which communication between those left behind and the dead remains possible. The terms used in the texts are the sumerograms ZI = Hitt. *ištan(zan)*[34]- and GIDIM = Hitt. *akkant-* lit. "dead". They sometimes seem to be used indiscriminately, but are certainly not synonymous. The ZI is any individual's, either dead or alive, seat of emotional and rational thoughts.[35] GIDIM is the "(ghost of the) dead", that may be invoked after death and through which contact can be established.[36] Their relation may be compared to that of soul and body before death, that is, the GIDIM may have been conceived of as more "corporeal" than the soul, as some immaterial but potentially visible body. In this respect one could recall the encounters of Odysseus and Aeneas in the Underworld with their mother and father, respectively. Seeing them, talking to them, they both want to embrace their parent but grab with their arms empty space.

Contact with the Netherworld could be established by way of magic ritual. During an evocation of the gods of the Netherworld, "[w]hen they clean a house from blood(shed), defilem[ent], injury? (and) perjury",[37] a Hittite priest descends to a riverbank, digs a pit and slaughters a lamb over it. He then invokes the Sungoddess of the Earth to admit the gods of the Netherworld to him. Later on, he digs another pit for these gods using a dagger or sword, pours a libation of oil, honey and wine in it and finally throws in one silver shekel, the "standard" means

34 Cf. N. Oettinger, "Die *n*-Stämme des Hethitischen und ihre indogermanischen Ausgangspunkte," *KZ* 94 (1980), 58-59.

35 See A. Kammenhuber, "Die hethitischen Vorstellungen von Seele und Leib, Herz und Leibesinnerem, Kopf und Person," *Zeitschrift für Assyriologie* 56 (1964), 160-162, and more in general ibid. 154-168.

36 See H. Otten, *HTR* 143-144.

37 KUB VII 41 i 1-2 (CTH 446, cf. H. Otten, "Eine Beschwörung der Unterirdischen aus Boğazköy," *ZA* 54 (1961), 116-117).

of payment. As has been rightly observed[38], this ritual shows striking resemblances with the prescriptions Odysseus received to enter into contact with the Netherworld in order to see the diviner Teiresias. In the process, he also meets many of his former combatants as well as his mother. It may be objected, however, that the contact in the Hittite ritual does not explicitly involve the *manes* of formerly living individuals but the deities of the Netherworld. On other occasions, however, contact with human forefathers was indeed established. Several oracle texts inform us about the anxiety certain crimes of the past aroused in later times: sometimes the spirits of persons involved were conjured up, and attempts at pacifying them were made.[39] Although the way this was done remains largely unknown, effigies may have played an important role: effigies were used as substitutes for living human beings in rituals in order to transfer some evil unto them, effigies of deceased kings and queens were the focus, as we will see, of both the ancestor cult and the royal funerary ritual, and effigies could be used in voodoo magic. A woman called Nikkalluzzi, for instance, tells how she tried to bewitch an enemy of the Hittite king:

> We made two effigies. One effig[y of cedar wood we made], the other effigy of clay we made. [On the effigy of] cedar wood we p[ut] the name of the enemy of the Majesty, while on the effigy of clay [we put] the name BU[-LUGAL]. Now, on/to the effigy of BU-LUGAL [...][40]

The Hittites certainly practised ancestor cult as far as the royal family was concerned. The renowned king lists, primarily used for their help in establishing the order and number of Hittite Great Kings and Queens, amply attest to this practice.[41] According to these and similar texts, statues of deceased kings had been erected in temples and received offerings. To what extent this ancestor cult was practised outside the royal family remains unknown, but the fact that it was done there fits in with the general belief that a Hittite king or queen as opposed to any ordinary mortal became a god after death. With few exceptions[42] the death of a

---

38 G. Steiner, "Die Unterweltsbeschwörung des Odysseus im Lichte hethitischer Texte," *Ugarit Forschungen* 1 (1969), 265-283.

39 Cf. G. del Monte, "Il terrore dei morti," *Annali dell'Istituto Universitario Orientale di Napoli* 33 (1973), 373-385, id., "La fame dei morti," *AION* 35 (1975), 319-346, and in *Archeologia dell'Inferno. L'Aldilà nel mondo antico vicino-orientale e classico*, ed. P. Xella (Verona 1987), 95-115, A. Archi, "Il dio Zawalli. Sul culto dei morti presso gli Ittiti," *AoF* 6 (1979), 81-94.

40 KUB VII 61, 4-8 (CTH 417, cf. M. Hutter, "Bemerkungen zur Verwendung magischer Rituale in mittelhethitischer Zeit," *AoF* 18 (1991), 39-40), compare also KUB XL 90 (Bo 1624) apud R. Werner, *Hethitische Gerichtsprotokolle* (Wiesbaden 1967; = *StBoT* 4), 64-65.

41 Cf. especially H. Otten, "Die hethitischen 'Königslisten' und die altorientalische Chronologie," *Mitteilungen der Deutschen Orientgesellschaft* 83 (1951), 47-71, id., *Die hethitischen historischen Quellen und die altorientalische Chronologie*, 103-106, 111 and 126.

42 Cf. e.g. KBo VI 29 i 22 (CTH 85, ed. A. Götze, *Ḫattušiliš. Der Bericht über seine Thronbesteigung nebst den Paralleltexten* (Leipzig 1925), 46-47) GIM-*an=ma* *ABU=IA* *kuwapi* BA.ÚŠ "When my father (i.e. king Muršili II) had died."

king was formulated as, for instance, "when my father became god." This difference in fate may be seen also in the substitution ritual quoted above. During the days on which the several necessary rites are performed, the king daily addresses himself in the early morning to the Sungod of Heaven with the following prayer:

> Sungod of Heaven, my Lord, what have I done that you have taken from me (my) th[rone] and given it to someone else? ... You have summoned me to the (ghosts of the) dead and, be[hold], (here) I am among the (ghosts of the) dead. I have shown myself to the Sungod of Heaven, my Lord, so let me ascend to my divine fate, to the gods of Heav[en] and [free] me from among the (ghosts of the) dead.[43]

By temporarily stepping down the king has become a mere mortal and already feels himself surrounded by the *manes* and begs to be admitted to kingship again, which promises him a place among the gods after death. But in spite of both the formula of "becoming a god" and the passage just quoted, it is rather surprising that the deceased kings and queens never receive the divine determinative (^{d}), as opposed, for instance, to the Roman emperors and some of their family members whose names, after having been awarded divine status by the senate through the act of *consecratio,* are forever accompanied by the adjective *divus*.

In the two passages quoted earlier in connection with the expression "the day of the mother", the Sungoddess of the Earth was mentioned, and there was reason to believe that the deceased king was thought to be in the Netherworld initially, from where he was taken away by his mother. When queen Puduḫepa asked for a long life for her husband, she addressed herself to the Queen of the Netherworld Lelwani. Moreover, an important part of the funerary ritual for a Hittite king consists of combined offerings to – among others – the Sungod of Heaven or the Sungoddess of Arinna as well as the Sungoddess of the Earth. It may be, as we already suggested, that offerings to the latter deity were meant to pacify the gods of the Netherworld in order to prevent their demanding the king's soul.[44] This hypothesis finds support in the anxiety the king expresses in his

43 KBo XV 2 (w. dupl.) rev. 14'-19' (CTH 421, ed. H.M. Kümmel, *StBoT* 3, 62-63).

44 E. Masson, *Les Douze Dieux*, 55-71, suggests that before ascending to the gods a king or queen after death first had to go to the Netherworld, that is, through a purificatory "passage sous la terre et par les eaux" in order to obtain immortality. Although the two passages quoted in connection with the expression "the day of the mother" seem to imply that the king first went into the Netherworld, from where his Mother would then bring him to heaven, the Hittite sources do not seem to support the idea of this stay as a rite of passage. The texts adduced by her all come from rituals that are meant to free individuals or houses from all kinds of impurities, none with a funerary character. The earth serves here as a place to safely store the evils, never to surface again.

prayer to the Sungod of Heaven just quoted.[45] Interesting at this point may be a short but enigmatic scene enacted on the second day of the royal funerary ritual:

> ... The 'Old Woman' takes a pair of scales. On one side she lays silver, gold and all (kinds of) precious stones, on the other, however, she lays clay.
> The 'Old Woman' speaks as follows to her colleague while mentioning the deceased's name: 'Will they bring him, so-and-so? Who will bring him?' Her colleague then answers: 'The men of Ḫatti (and?) the *uruḫḫi*-men will bring him.' And the first one says: 'Let them not bring him!' but her colleague says: 'Take the [si]lver (and) gold.' And the first one says: 'I will not [ta]ke it' and she says so three times.
> The third time the first speaks thus: 'The clay I will ta[ke', and] she breaks the pair of sca[les]. She [holds?] them up tow[ards] the Sungod [and ... si]ngs? and [starts] to wail.[46]

O.R. Gurney[47] tries to explain the scene as a "market transaction" in which the deceased is symbolized by the earth or clay and his weight in gold, silver and precious stones is offered to the mysterious *uruḫḫi*-men: "but their price is refused, he (i.e. the deceased) is redeemed." If, however, they are the ones bringing him, he cannot be offered for sale to them. Moreover, in this interpretation the *uruḫḫi*-men, whether equivalent to the Ḫatti-men or in cooperation with them, represent the powers of the Netherworld, which seems a very unexpected role for the king's own countrymen. Finally, if the "Old Woman" wants to free the deceased from the *uruḫḫi*-men, then her urgent appeal not to bring him, does not make much sense. Could it be that the second priestess plays the part of the Netherworld, trying to lure her colleague into taking the valuables thereby leaving the deceased to her? In such an interpretation the 'Old Woman' may not even have wanted the deceased to be present in such a potentially perilous conversation. In order to prove that the sale was not made, she then shows the broken scales to the Sungod of Heaven.

In spite, then, of the ultimately divine fate, which the members of the royal family after death enjoyed, the Netherworld and its gods may have initially housed

---

45 Taking up a suggestion by J. de Kuyper, "Les Obligations Funéraires d'une Reine Hittite," *Studi Epigrafici e Linguistici* 4 (1987), 97-98, one could see the same concern to be freed from the Netherworld in Ḫattušili's I words at the very end of his so-called 'Testament': *nu=mu ... taknaz paḫši* "protect me from the earth!"; for this much disputed passage see St. de Martino, "Hattušili e Haštayar: un Problema Aperto," *Oriens Antiquus* 28 (1989), 1-24 with all relevant literature, and H.C. Melchert, "Death and the Hittite King," in *Perspectives on Indo-European Language, Culture and Religion. Studies in Honor of Edgar C. Polomé*, (McLean, Virginia 1989), 182-188.

46 KUB XXX 15+ obv. 26-36 (CTH 450, ed. H. Otten, *HTR* 68-69); the translation "clay" in line 34 is based on the liturgy text KUB XXXIX 41 obv. 13' (cf. H. Otten, *OLZ* 57 (1962), 231-232).

47 *Schweich* 60.

them, and there is some support for this notion in archaeological evidence as we will presently see.

TEXTS AND ARCHAEOLOGY: SOME POSSIBLE FUNERARY STRUCTURES

In the Hittite texts there occur three architectural terms designating buildings with a clear funerary purpose. These are the É.NA$_4$ (DINGIR-*LIM*) or "Stone House (of the god)/(divine) Stone House", the ^{NA4}*ḫekur* (SAG.UŠ)-house and the É.GIDIM (or É *ŠA* GIDIM), "House of the (ghost of the) dead."[48] In all three cases we are dealing with places where either the remains of the king had been deposited or where a special cult for a deceased king was situated. These entities did not necessarily consist only of buildings or monuments, but presumably all were institutions comprising grounds, properties and personnel. Among the latter were farmers tilling the land belonging to the institution, herdsmen and craftsmen. Probably, such a community was self-supporting to a high degree and rather reclusive as well. This may be deduced from a stipulation concerning a "Stone House" according to which men within the community could marry women from outside on the condition that they would join the community; no one, however, was permitted to leave for a marriage outside. The same document illustrates the importance attached to the cult by the fact that it granted freedom from taxes to the "Stone House."[49] This last regulation was not exceptional and is attested for the other two institutions as well.[50]

The É.NA$_4$ or "Stone House" may have been a real mausoleum: the place where the remains of the deceased find their ultimate resting place. In the royal funerary ritual the ashes and bones are deposited there on the second day after having been burned on the pyre:

> [The ...] have brought from the palace two oxen (and) two times nine sheep. One [ox (and) nine sheep] they dedicate to the Sungoddess [of the Earth], the other ox and nine sheep [they dedicate to the soul of the] deceased. [... the b]ones they pick up and from the pyre they [...] them and bring [them] into his Stone-house. In the inn[ermost] of the Stone-house they set up [a be]d, t[ake] away the bones from the throne [and] place them on the bed, that was set up. A lam[p weighing] ... shekels with fine oil they

48 The link between the É.NA$_4$ and the *ḫešta*-house hinges on the identification — whether by real or folk etymology — with the É *ḫaštiiaš* "house of bones" (Apology of Ḫattušili III, iv 75). Obvious at any rate is the close connection of the *ḫešta*-house with the cult of Lelwani, the Netherworld deity, and chthonic gods in general. Conclusive evidence for its use as a real mausoleum or as a temple dedicated to the ancestor cult seems to be lacking; cf. O.R. Gurney, *Schweich* 38-43, V. Haas – M. Wäfler, *UF* 8 (1976), 65-99 and ibid. 9 (1977), 87-122, and J. Puhvel, *Hittite Etymological Dictionary* (Berlin – New York 1984ff.), 3: 319-323.

49 KUB XIII 8//KUB LVII 46 (CTH 252, cf. H. Otten, *HTR* 104-107).

50 See F. Imparati, "Le Istituzione Cultuali del ^{NA4} *ḫékur* e il Potere Centrale Ittita," *Studi Micenei ed Egeo-Anatolici* 18 (1977) 19-64 for the *ḫekur*-house, for the É (*ŠA*) GIDIM cf. ABoT 56 iii 4'-5' with Otten's restoration *HTR* 104 after KUB XIII 8 obv. 6.

> put in front of the bones [... . The]n they dedic[ate] one ox and one sheep to the soul of the deceased.[51]

Later on, several objects of gold, silver and other precious materials that were used during the various rites are brought into the "Stone House" as well.

The evidence for the *ḫekur*-house, however, hints at a place which could be dedicated to the cult of a deceased king or queen without necessarily containing their remains. The last known Hittite king, Šuppiluliyama (II), tells how he made a statue of his father and described in detail his *res gestae* and then continues: "An everlasting *ḫekur* I built. A statue I made and brought it into the everlasting *ḫekur*. I put it in its proper place (and) dedicated(?) it."[52] The statue of a queen in connection with a *ḫekur* is mentioned in an oracle text.[53] Here it seems to be the living queen who is preparing her own memorial. She decorated it, however, with a gold wreath which she denied the deity of the city of Arušna, whose wrath she thereby brought on herself. Hiding in her avarice the gold wreath and trying to satisfy the deity with two silver ones instead, the queen only made matters worse. She was temporarily banned from the palace until in a letter to the king she disclosed the hiding place. Trouble involving another queen and both an É ^{NA4}*ḫekur* ^{d}LAMMA "*ḫekur*-house of the Tutelary Deity" and an É.NA$_4$ DINGIR-*LIM* "divine Stone-house/Stone-house of the god" is attested to in the affair concerning Muršili II's stepmother, the last wife of his late father. In an emotional appeal to the gods Muršili says:

> You, o Gods, don't you see, how she has turned over my father's complete estate to the *ḫekur*-house of the Tutelary Deity (and) the divine Stone-house? One thing she had brought over from the town of Šanḫara, another she gave away to the entire population in Ḫattuša. Nothing she left behind! Don't you Gods see?[54]

The juxtaposition of the two terms in this passage indicates how closely interrelated they were without, however, being identical. Very often these *ḫekur*-houses are accompanied in the texts by either a geographical, divine, or some other (mostly obscure) name. From the geographical names it may be inferred that they could be located outside Ḫattuša. This possibility seems to be discussed in the royal funerary ritual in a fragment probably belonging to the description of the events of the second day:[55]

---

51 XXXIX 11(+) 44'-51' (CTH 450, ed. H. Otten, *HTR* 68-69).

52 KBo XII 38 ii 17'-21' (CTH 121, ed. H.G. Güterbock, "The Hittite Conquest of Cyprus Reconsidered," *Journal of Near Eastern Studies* 26 (1967), 76, 78).

53 KUB XXII 70 obv. 12-27 (CTH 566, ed. A. Ünal, *Ein Orakeltext über die Intrigen am hethitischen Hof (KUB XXII 70 = Bo 2011)* (Heidelberg 1978; = *THeth.* 6), 6-61).

54 KUB XIV 4 ii 3'-8' (CTH 70, cf. CHD L-N 361b with literature).

55 KUB XXXIX 12 (for the possible indirect join to KUB XXX 15+KUB XXXIX 19(+)11, see H. Otten, *HTR* 64)//KUB XII 48//KBo XXXIV 55 (CTH 450).

> [The ...] they will pick up, [but if in another city the body has been b]urned (or: [to another city the bones? have been b]rought), in that [city ...] they will pick them up, and in which city there (is) a 'Stone house' for him, [to that city they will b]ring? [them], until, however, (his) bones in/to that city [...], they will [...]. Finally?, in/to which city [...], if there in that city there (is) his 'Stone-house', then there to that city they will transport them (i.e. the bones).[56]

One of these *ḫekur*-buildings in the countryside took on considerable political importance, as it seems. Located somewhere in southern Anatolia, it was probably the cult place for king Muwatalli II, who died around 1274 BC. His eldest son and successor was deposed and subsequently banished by his uncle. In order to keep the youngest son from honoring his father and thereby displaying his legal rights to the throne Ḫattušili III, the usurper king, and initially also his son Tudḫaliya IV denied him access to the shrine.[57]

Although no obvious royal tombs have been found so far, at least two sites in and around the capital Ḫattuša come into consideration for having been funerary monuments of the kind discussed here (see Fig. 1).[58] In the text KBo XII 38 just quoted Šuppiluliyama (II) recorded the dedication of an "eternal *ḫekur*" to his father Tudḫaliya IV. It has already been pointed out by others[59] that the cuneiform text is the translation of a (monumental) hieroglyphic inscription and that this might be the very much weathered and barely legible inscription on the stony outcrop of Nişantaş in the Upper City of Ḫattuša. Up to now, only the beginning of this text had been definitively interpreted and seemed to match the cuneiform version. Recent archaeological research of the site now tends to confirm this, and as a consequence, Nişantaş may be considered a *ḫekur*-structure.[60] The building on top of this outcrop measures 38x35 m with a monumental entrance gate flanked by sphinxes reached by a slowly ascending slope of approximately 35 m. The scanty remains of this building, however, do not allow us to reconstruct anything more than the ground plan of the building, nor were any finds recorded in support of the supposed cult purpose. We cannot, therefore, conclude that we have found the grave of Tudḫaliya IV. First of all, if on the combined evidence of the cuneiform and hieroglyphic text this was indeed the *ḫekur* for Great King Tudḫaliya

---

56 KUB XII 48 obv. 2'-8'//KUB XXXIX 12 rev.? 2'-6'//KBo XXXIV 55, 1'-3' (CTH 450, ed. H. Otten, *HTR* 70-73).

57 Cf. Bo 86/299 (Bronze Tablet) i 91-ii 3; ed. H. Otten, *Die Bronzetafel aus Boğazköy. Ein Staatsvertrag Tudḫaliya IV.* (Wiesbaden 1988; = *StBoT Bh.* 1), 14-15 and 42-45, Ph.H.J. Houwink ten Cate, *ZA* 82 (1992), 244-249.

58 According to K. Bittel, *Die Hethiter* (München 1976), 114, the stony outcrops in Boğazköy of Sarıkale and Yenicekale, and of Gavur Kalesi (some 60 kms. south-west of Ankara) may also be considered *ḫekur*-structures.

59 H.G. Güterbock, *JNES* 26 (1967), 74, 81, E. Laroche, *Anatolica* 3 (1969-1970), 93-98 with Pl. V-VII.

60 Cf. P. Neve, "Die Ausgrabungen in Boğazköy-Ḫattuša 1991," *Archäologische Anzeiger* 1992, 323-333.

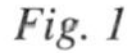

Fig. 1

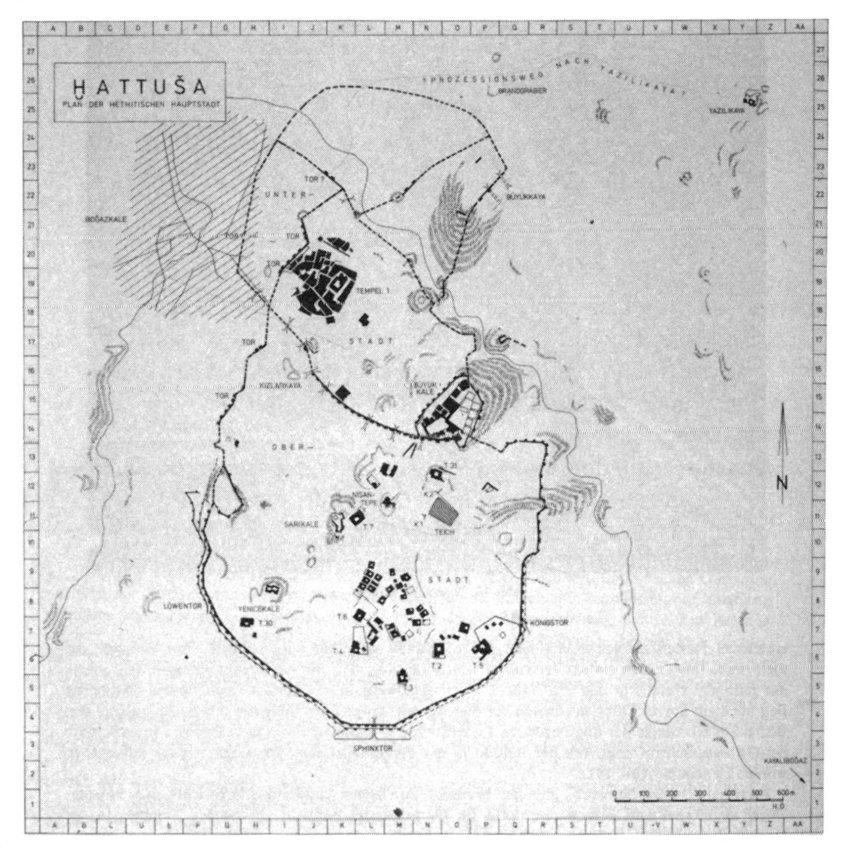

Map of the ancient Hittite capital Ḫattuša, nowadays Boğazköy (Boğazkale) (courtesy Dr.-Ing. P. Neve, source AA 1993, 622)

IV, his son built it probably years after his father's death, certainly not on that occasion. Moreover, he speaks of a statue and says nothing about transferring his father's ashes from somewhere else. Secondly, we have to deal with the funerary character of the so-called Room B in the rock sanctuary of Yazılıkaya immediately outside Ḫattuša with its clear dedication to Tudḫaliya IV. This room – almost 20 m long but narrow, only a couple of meters wide, adjacent to the main sanctuary (room A) and connected with it through an even narrower passageway[61] – bears the relief (nr. 82; see fig. 2) of a deity's head as the top of a dagger stuck into the ground with a hilt ornamented with lions. This so-called Sword God can be linked by textual evidence to the Netherworld as well as the relief (nrs. 69-80)

---

61 For a description of this room see R. Naumann in K. Bittel et al., *Das hethitische Felsheiligtum Yazılıkaya* (Berlin 1975), 39-49, and K. Bittel, ibid. 158-165; see also P. Neve in *Anatolia and the Ancient Near East*, ed. K. Emre, B. Hrouda, M. Mellink, N. Özgüç (Ankara 1989), 345-355 with Pl. 65.

Fig. 2

Yazılıkaya Room B eastern wall with reliefs of the Sword God, Great King Tudḫaliya IV in the embrace of his protective deity Šarruma and one of the niches (Photo courtesy Dr. D.J.W. Meijer).

of twelve marching gods on the opposite wall (see fig. 3).[62] Furthermore, there is the large relief (nr. 81) of the king embraced by his patron deity Šarruma (see fig. 2) and a smaller cartouche (nr. 83) with his name in hieroglyphic script usually taken to refer to a colossal statue at the northern end of the room of which only the pedestal with the imprint of two feet remains. All relief figures are oriented towards this end of the room. Finally, along the two long walls there are three niches (see fig. 2) that may once have contained the ashes of the Great King and, possibly, his immediate family.[63]

Thus, room B of Yazılıkaya displays most obviously a funerary character because of the reliefs and – if interpreted correctly[64] – the columbaria, but referring to the Netherworld[65] and not so much to the apotheosis of the king. In the case of Nişantaş it was the combination of the cuneiform and the hieroglyphic text that led us to believe that we are dealing with a genuine *ḫekur*-monument. If Yazılıkaya Room B was once the real 'grave' or mausoleum, then the more temple-like structure of Nişantaş may have been meant to serve the cult of the deceased king in his deified form. In other words, Room B may have been dedicated to king Tudḫaliya's "body natural", Nişantaş to his "body politic."[66]

---

62 For a description of this relief see K. Bittel, *Yaz.*[2] 163-164; for the textual link with the Netherworld see H. Otten, *ZA* 54 (1961), 122-123 with commentary on pp. 148-149.

63 See for this possibility R. Naumann, *Yaz.*[2] 45, and K. Bittel, ibid. 255-256.

64 Compare the sceptic remarks of O.R. Gurney, *Schweich* 63, and P. Neve, *FsÖzgüç* 351-352.

65 Compare also the remarks of P. Neve, *FsÖzgüç* 349.

66 Such a distinction brings to mind the Roman usage of the *funus in corpore* and *funus in effigie* (see also below n. 99); compare the remarks by E. Bickermann, "Die römische Kaiserapotheose," *Archiv für Religionswissenschaft* 27 (1929), 23: "Die römische Religion scheidet ... haarscharf zwischen dem verewigten *divus* und dem toten Kaiser, dessen Leichenreste im Mausoleum beigesetzt sind. (...) Die Grabsteine der kaiserlichen Mausoleen erwähnen nie die Gotteswürde des Konsekrierten, die Denkmäler der *divi* verschmähen ihrerseits die irdischen Dignitäten zu nennen."

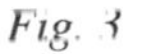

Yazılıkaya Room B western wall with relief of the twelve marching gods (Photo courtesy Dr. D.J.W. Meijer).

HITTITE CEMETERIES: THE ARCHAEOLOGICAL RECORD

Up to now all texts and even the archaeological material at our disposal referred to the king or the royal family, the archives being those of the palace or temples and the country being largely administered by the royal family in its extended sense. As a consequence, the beliefs of the population concerning death remain largely unknown as far as the texts are concerned. Fortunately, however, archaeology is able to fill this gap to a certain extent. Several Hittite cemeteries are known (see fig. 4): Osmankayası and Bağlarbaşıkayası[67] directly north of the

Fig. 4

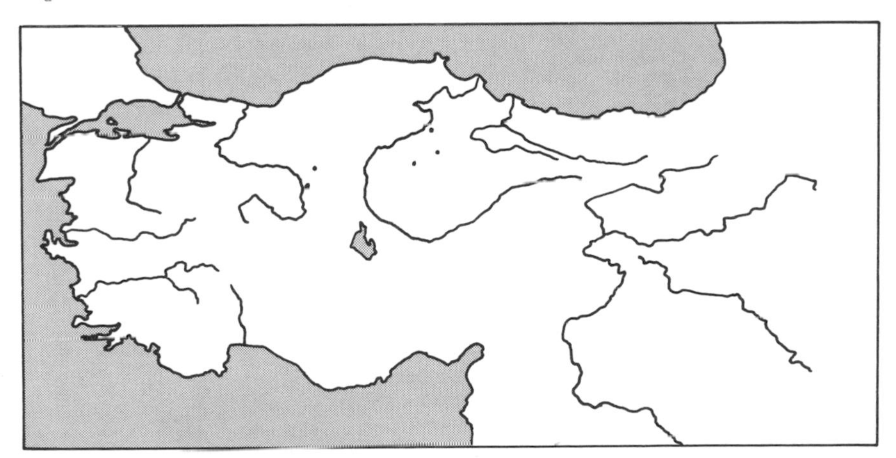

Map of Turkey with cemetery sites.

67 K. Bittel et al., *Die hethitischen Grabfunde von Osmankayası* (Berlin 1958).

capital Ḫattuša and probably its extramural burial grounds, Gordion[68] some 200 km west of Ankara, and Ilıca[69] and Kazankaya[70] north of Boğazköy/Ḫattuša being the most important.[71] In all places the graves are mostly to be dated to the first half of the second millennium, but there is third millennium material as well. The burial methods differ considerably among the sites mentioned. Osmankayası and Bağlarbaşıkayası show a seemingly contemporaneous mixture of cremation and inhumation, Gordion shows both types as well, but cremation was not introduced before the second millennium. At Kazankaya only inhumation is attested, at Ilıca only cremation except for one cist grave. Regardless of the burial method, all graves tend to be very sober.[72] On the whole there are few gifts: some rings, pins that may have held clothing, spindles and/or spindle whorls, and (simple) ceramic pottery. At Gordion also earrings, beads and necklaces were found. The Gordion and Ilıca sites were both found in a relatively undisturbed situation and so offer the best view of Hittite burial practices. The inhumation graves at Gordion held the bodies mostly in contracted position in cists or pithoi of ca. 1 m. in height and covered by a slab of stone or a ceramic lid. Only the oldest third millennium graves lack cists or pithoi. In the cist graves there does not seem to have been a preference for laying the body on either the right or left side. There were no family graves. Along with the introduction of cremation in the second millennium, however, extended burials came into use in the so-called 'double pithoi-graves.' At Ilıca, where with one exception only cremation was practised, a burial pit was dug as big as the vase it was to contain, mostly with a depth of 0.6-0.8 m. All vases here are of the *Schnabelkanne* type. These vases were then put in the pit supported by stones to secure their upright position.

The situation at Osmankayası is somewhat different. The cemetery there was originally formed by an overhanging rock creating a sort of cave. The cremation practice largely resembles that of Ilıca, although the range of vase types at Osmankayası is larger by far, and there do not appear to have been neatly dug pits to hold them. The pottery is of the normal type, that is, no special funerary ceramics seem to have been in use. Remarkable, though, is the observation made at Osmankayası and Ilıca alike that almost all the pottery found was already damaged on purpose in ancient times. Bittel, the excavator of the former site, has suggested the pots were broken so as not to be used anymore by the living: "um für immer den Toten zu gehören", but this may also be interpreted as an expression of

---

68 M. Mellink, *A Hittite Cemetery at Gordion* (Philadelphia 1956).

69 W. Orthmann, *Das Gräberfeld bei Ilıca* (Wiesbaden 1967).

70 T. Özgüç, "A Hittite Cemetery at Kazankaya" in *Excavations at Maşat Höyük and Investigations in its Vicinity* I 69-88; in the same volume Özgüç reports about an unidentified and plundered cemetery near the village of Büget (8 km north of Çorum).

71 For a more detailed list with bibliography see W. Orthmann, *Reallexikon der Assyriologie und Vorderasiatischen Archäologie*, ed. D.O. Edzard (Berlin – New York), 3: 605 s.v. "Grab", and M. Marazzi, *L'Anatolia Hittita. Repertori archeologici ed epigrafici* (Roma 1986), 52-57.

72 Although the cemetery at Kazankaya was found plundered and as a consequence cannot be used as an argument in this respect, it is not likely to have differed much from the other sites discussed here.

grief.[73] At the same two burial sites vases sometimes showed additional holes around the neck or in the bottom. At Osmankayası often more individuals seem to have been cremated at the same time, adults and children. The burial site was certainly not the place of cremation: where this was done is not known. As far as inhumation is concerned, bodies were buried in a contracted position. Sometimes complete animals were buried alongside: oxen, pigs, sheep and goats, or equids, mostly mules; often only their heads are found (cf. below, page 68). The equids were all males at the height of their strength. Although Osmankayası is not said to have been secondarily disturbed, there are hardly intact burials. This is due to the fact that in burying recently deceased individuals, older graves seem to have been simply pushed aside without any apparent respect. Exception was only made for the skulls, which were gathered and 'buried' again.

Paleo-anthropological research on the human skeletal remains carried out most extensively at Osmankayası[74] presents us with some shocking results. The predominant age group is between 14 and 20 years old. This more or less matches the results reached at Ilıca, where no individuals were found older than 30-35 years. Dental research testifies to primitive forms of food preparation. Interesting is the degree of caries in comparison with material from Europe in the same period. In general, caries was encountered more often and worse than in Europe, older remains showing a lesser degree than younger ones, though. According to the paleo-anthropologist J. Schaeuble this could point to a somewhat more advanced degree of civilization in Anatolia at the time as compared with Europe.

On the basis of the ceramic material the cemetery of Osmankayası must have been in use at least until the 14th century. The written sources of the 14th and especially those of the 13th century regularly attest to the fact that people of the highest social level could live considerably longer than 30-35 years. King Ḫattušili III, mentioned before, lived to be older than 70, his wife Puduḫepa must have been ca. 90 years old when she died, and there is evidence for several high functionaries reaching ages between 60 and 80. It has been suggested[75] that because of the extraordinarily simple burials, the graves at Osmankayası must have belonged to the poorest in Hittite society. This may partly be true, and it can at least be stated with some probability that those pertaining to the highest social levels were not buried there. On the other hand, as no significantly different burials have been found either at Boğazköy/Ḫattuša or anywhere else in Anatolia in the same period, the possibility must be taken into consideration that we could also be dealing with a custom of simple burial. It may be that whatever the popu-

---

73 One might point to the so-called destructive sacrifices known from ancient Greece, cf. W. Burkert, *Greek Religion* (Oxford 1985), 192-193: "In addition to grave offerings, there are destructive sacrifices, motivated by the helpless rage which accompanies grief: if the loved one is dead, then all else must be destroyed as well. Weapons and tools are broken; dogs and horses, and even the servants and wife of the dead man may be killed."

74 J. Schaeuble, "Anthropologische Untersuchung der hethitischen Skelettfunde aus Osmankayası bei Boğazköy" in K. Bittel et al., *Die hethitischen Grabfunde von Osmankayası* , 35-59.

75 K. Bittel, ibid. 23-24.

lar belief in an afterlife was, if any, no real importance was attached to a rich burial. Practical considerations like the chance of the grave being plundered may have played a role in such matters.[76]

For the 13th century one could say that the royal family in the largest sense of the word, i.e. including numerous sidelines and those married into this family, held all key positions in the administration, whether military or civil. Thus, it may be that many of those were finally buried or their ashes deposited in royal mausolea and are therefore not to be found at places like the cemeteries mentioned here.

THE HITTITE ROYAL FUNERAL: INTRODUCTION

Several times already we have referred to the Hittite royal funerary ritual. The text ensemble as listed by E. Laroche in his Catalogue des Textes Hittites under nr. 450 describes a ceremony for a deceased king or queen. A similar ritual must have existed for their children.[77] The only reference to this royal ceremony outside this corpus provides the Hittite term for such rituals: *akkantaš šaklaeš* "rites (*šaklaeš*) for the dead."[78] It can be found in the extensive Annals of king Muršili II (1322-1295 BC), where he describes the death of his brother Šarrikušuḫ, king of Kargamiš: "Šarrikušuḫ, [m]y [brother ... fe]ll ill and died. They [b]rought him to Ḫattuša and such as are the rit[es] for the dead, [these] they performed in Ḫattuša."[79]

The title of the composition describing these rites used by the Hittite scribes in the colophons and by which it was catalogued in their archives was the incipit: *mān* URU*Ḫattuši šalliš waštaiš kišari* "When in Ḫattuša a great loss occurs." Traditionally, the Hittite term *uaštaiš* is translated as "sin, shortcoming." This word derives from the verb *wašt*(*a*)- the basic meaning of which seems to be something like "to deviate" resulting on the one hand in the most frequently occurring sense "to sin, err", and on the other hand in the sense "to miss (a

76 One is reminded here of the first millennium Aramaic inscription of the priest Si'-gabbari; there we read: "(...) They (i.e. his children and their offspring) did not lay with me any vessel of silver or bronze; with my garments (only) they laid me, so that in future my grave should not be dragged away (...)", cf. J.D. Hawkins, "Late Hittite Funerary Monuments," *Mesopotamia* 8, 216.

77 Cf. KUB XXXIX 6 iii 14-15 (outline, CTH 450, cf. H. Otten, *HTR* 50-51): *mān* DUMU.NITA - *ma našma* DUMU.MUNUS DINGIR-*LIM*-*iš* / *kišari* "If a boy or a girl becomes a god." The fact that this entry immediately follows the outline for the royal ritual as well as the expression that the child "becomes a god", make it sufficiently clear we are dealing with a similar ritual for princes and princesses.

78 This combination is written *ŠA* GIDIM.ḪI.A (...) *šaklaeš* and can be read only by combining the duplicating texts: KUB XIV 29 i 29 (*ŠA* GIDIM.Ḫ[I.A), KBo IV 4 i 8 (*ša-ak-l*[*a*-) and KBo X 38, 7' (*Š*]*A* GIDIM.ḪI.A ... *š*[*a*-; for this last duplicate cf. P. Meriggi, "Spigolando nei Testi Storici Etei", *Orientis Antiqui Collectio* 13 (1978), 67-70, reference courtesy of Ph.H.J. Houwink ten Cate); for the equation Sum. GIDIM = Hitt. *akkant*- see H. Otten, *HTR* 143-144, and HW^2 A s.v. *akkant*-.

79 KBo IV 4 i 5'-8'//KUB XIV 29 i 28'-30'//KBo X 38, 2'-8' (CTH 61, ed. A. Götze, *Die Annalen des Muršiliš* (Leipzig 1933), 108-109).

target)."[80] The adjective *šalli-* "great" may have the connotation of "royal": this is, for instance, the case in the well-known expression for "throne, kingship"*šalli pedan*, lit. "the great place."[81] Similarly, the combination *šalli ḫaššatar* "great/extended family" is used in reference to the royal family exclusively. Thus, the above title of the composition may to Hittite ears have clearly referred to royal death.

The text material for the funerary ritual now available comprises some 100 larger and smaller fragments, which in many cases have been joined in the past to fairly large portions of once complete tablets. Many gaps, moreover, could be filled by the identification of duplicates due to the Hittite custom of copying texts for their archives and libraries. We often have three or even four versions of one text, sometimes exactly matching each other or showing only minor differences, sometimes, however, differing up to the point where it is better to speak of 'parallel versions' rather than duplicates. Almost all manuscripts were written in the 13th century; two fragments date back to the 14th century or earlier.[82] The language itself also makes it clear that the composition probably goes back to middle (ca. 1450-1350) or even old Hittite (ca. 1650-1450) times.[83]

---

80 For a detailed semantic and phonological analysis of the root underlying this verb see J. Catsanicos, *Recherches sur le Vocabulaire de la Faute. Apports du Hittite à l'étude de la phraséologie indo-européenne* (Paris 1991); for a different opinion see J. Puhvel, "Shaft-shedding Artemis and mind-voiding Ate: Hittite determinants of Greek etyma," *Historische Sprachforschung* 105 (1992), 4-8.

81 Its antonym *tepu pedan* litt. "the small/humble place" (cf. H. Otten, *HTR* 128) is attested in passages which may have to do with loss of kingship or maybe even with a king's demise. Besides one occurrence in the funerary ritual itself (KUB XXX 25+ rev. 10, cf. H. Otten, *HTR* 28-29) we find this combination in the myth of Ullikummi at the point where the reigning Stormgod has lost his first battle against his contenders. Although the passage there (KUB XXXIII 106 ii 5, cf. H.G. Güterbock, "The Song of Ullikummi. Revised Text of the Hittite Version of a Hurrian Myth," *Journal of Cuneiform Studies* 6 (1952), 21-21) is somewhat damaged, it seems to convey that he temporarily quits the throne (so H.A. Hoffner, *Hittite Myths* (Atlanta, Georgia 1990), 58 and 61 n. 23). Another interesting example can be found in the list of tutelary deities KUB II 1 ii 39-40 *ŠA Labarna* / [^{d}LAM]MA *tepauwaš pē<d>aš lamarḫandattieš* (cf. G. McMahon, *The Hittite State Cult of the Tutelary Deities* (Chicago 1991), 102-103 with n. 94. Instead of the nom.pl. suggested by the CHD L-N s.v. *lamarḫandatt-* a nom.sg. of an *i*-stem *lamarḫandatti-* would be equally possible (although the problem of taking it as an apposition to ^{d}LAMMA remains), which in view of the similar passage ibid. iii 46 must mean something like "Tutelary deity of the Labarna fixing the time of the 'small place' ". Is this meant as fixing the time of the end of the king's reign/life? The last attestation comes from the prayer to the Sungod (CTH 372) in a passage parallel to the well-known phrase about life being bound up with death and a man's days being counted. The fragmentary state of the tablet ((KUB XXXI 127+) KUB XXXVI 79+ ii 46) only allows us to read *tepu pēdan* x[, but a reference to royal death might certainly be appropriate there. The text 660/c mentioned by Otten, *HTR* 128, as yet another instance of this combination is still unpublished.

82 XXX 17 (CTH 450 III 2; cf. H. Otten, *HTR* 50-53) and XXXIX 2 (CTH 450 I B; cf. H. Otten, *HTR* 18); XXXIX 64 is mentioned as older and belonging to funerary texts in a more general way by Otten, OLZ 57 (1962), 232-233, Laroche, CTH, however, attributes it to the festivals (CTH 670).

83 Cf. already H.C. Melchert, *Ablative and Instrumental in Hittite* (diss., Harvard University, Cambridge, Massachusetts 1977), 78.

The purpose of the composition is that of prescribing and its goals are wholly practical: when a king died, the organization for a complex 14-day funerary ritual had to be set in motion immediately. This ritual involved many individuals, over a hundred animals, huge quantities of foods and all sorts of objects that had to be ready for their role on a certain day at a certain time. And, as an important aspect of all rituals, it had to be performed impeccably: the text itself was not 'sacred', but its object most certainly was. The existence of copies and the practical character make it abundantly clear that we are not dealing with a description of a death ritual once carried out for a specific Hittite king, but with the traditional protocol of the Hittite royal funerary ritual. Because of its prescriptive character, the text is not meant to explain any of the symbolism or meaning of the rites: they were supposed to be known to the people involved. As such this composition is a unique document in the literature of the Ancient Near East comparable only to much later works as, for instance, *De ceremoniis aulae Byzantinae* by the Byzantine scholar emperor Constantine VII Porphyrogenitus (913-959) or to even later West-European treatises such as *De exequiis regalibus* (England, 14th century)[84], or *The manner of the ordering of the setting forth of a corpse of what estate that he be and of how every shall go in order after the estate and degree that he be of* (England, 16th century)[85]. These instructions were, of course, founded on experience, and officials in charge of the burials drew heavily upon the descriptions made of earlier ceremonies. This may very well have been the case with Hittite royal funerals, but we only have what we may then call the end product: the instruction for a fixed ceremony.

The practical requirements of the composition are neatly reflected in the organization of the material. Within the total of texts just mentioned the following series can be discerned[86]:

— the main (and often detailed) description of all ritual acts, and persons and objects involved therein
— the liturgy (*Rollenbuch*): the lines which the various 'actors' have to say on several occasions during the ritual
— the outline: an excerpt of the ritual according to the order of days, in which each day is characterized by a single sentence, maybe made for archival purposes
— the ration list (*Lieferungliste*): a detailed listing of all 'stage props' required per day, probably made up for the logistics of the ritual.

The same practical set-up was used for the various festivals of the Hittite cult calendar.

On the basis of the main description series the ritual performed on days 1-2, 4, 7-8-9 and 12-13 can be partly or completely reconstructed, that is more or less half

84 R.A. Giesey, *The Royal Funeral Ceremony in Renaissance France* (Genève 1960), 82-84.

85 P.S. Fritz, "From 'Public' to 'Private': The Royal Funerals in England, 1500-1830" in *Mirrors of Mortality. Studies in the Social History of Death,* ed. J. Whaley (London 1981), 76.

86 The terminology used for designating the four different series is borrowed from I. Singer in his edition of the KI.LAM-festival; cf. Singer, *The Hittite KI.LAM Festival. Part One* (Wiesbaden 1983; = *StBoT* 27), 13.

of the total number of days. For days 5-6, 10-11 and 14, the description can be supplemented by only (fragments of) short indications in the ration, outline and/or liturgy series. Fragments of the third day seem to be completely lacking in our records or have not yet been recognized as such.

THE HITTITE ROYAL FUNERAL: SUMMARY

The complete ritual lasted for 14 days. On the first day while the body lay in state, offerings were brought:

> When in Ḫattuša a great loss occurs, (that is,) either the king or queen becomes god, all, big and small, take away their reeds/straws and start to wail.
> On the day, that he/she becomes god, they do as follows. They dedicate one plow ox of the finest quality to his/her soul.
> They slaughter it at his/her head and speak thus: 'As you have become, let this one become likewise, and let your soul descend in this ox.'
> Then they bring a jug of wine and libate it to his soul, then they break it.
> When it gets dark, they swing one billy goat over the deceased thereby speaking thus: (break)
> Then [they give] him to drin[k ...] When with/out of a silv[er] cup [...]
> Another cup, however, of cl[ay ...] they put and [... they call the deceased?] by the name [...] When to him/her [...], then in front of the t[able ...] and on [the table? ...] and to the deceased [they ...] it.
> As long as they dr[ink ...], he d[rinks ...] as well.
> On that [day?] they keep [... -]ing and [they] stay awake during the night [?] The first day [ends here?].[87]

The expression "all (...) take away their reeds/straws" is still without parallel.[88] The Akkadian word used here (*šulpu* "reed") can designate either reed(s) as a musical instrument or a drinking straw. Is this to be taken metaphorically for refraining from all expressions of joy? Note how the deceased is still treated as if alive: he is called, and food and drinks are served to him.

In probably the same night, the body of the deceased was then burned. At dawn the next morning, the remains consisting of ashes and bones that had not been fully incinerated were gathered by women. This is the famous passage recalling the Homeric description of both Patroclos' and Hector's funeral in the Iliad and deserves to be quoted in extenso:

> When on the second day it becomes light, women go to the py[re] for the gathering of the bones and they extinguish the fire with ten jugs of beer, ten [jugs of wine] (and) ten jugs of *walḫi*-drink.

---

87 KUB XXX 16+KUB XXXIX 1 i 1-18, ii 1-14 (CTH 450, ed. H. Otten, *HTR* 18-21).

88 Cf. H. Otten, *HTR* 120-121.

> One silver *ḫuppar*-vase of half a mina and twenty <...?[89]> has been filled [wi]th fine oil. With silver tongs they take the bones one by one, and lay them in the fine oil in the silver *ḫuppar.* Then, they lift them out of the fine oil and lay them down on a *kazzarnuli*-cloth while under the cloth lies a fine cloth.
>
> When they finish gathering the bones, they wrap them up with the cloth and the fine cloth and put them on a throne to sit on; if it is a woman, however, they put them on a *ḫapšalli*-stool.
>
> Around that pyre on which the deceased has been burnt, they lay down twelve thick breads. On top of the thick breads they put oil cakes. The fire has already been extinguished with beer (and) wine. They put a table down in front of that throne on which the bones are lying and they give warm breads, GÚG-breads and sweet breads to break. Then the cooks (and) waiters place bowls at the head (of the table) and pick (them) up at the head (of the table). Each of all those who have come for the gathering of the bones, they give to eat.[90]
>
> Three times they give to drink and three times they drink his soul.[91] Thick breads (and) an Ištar-instrument are not involved. In front of the pyre (the action) has ended.
>
> [Aft]erwards they do as follows: In the mi[ddl]e of the pyre they form something of an image of a man (or?) a woman? wit[h fi]g(s), raisin(s) (and) olive(s). In its centre they lay fruit, *parḫuena* (and) *galaktar* of the gods, a right thigh bone (and) a flock of sheep's wool.
>
> In the middle of the image they pour beer, then in the middle they place thi[ck b]read(s) of three *sutu*'s (weight) ... [92]

Then follows the scene with the scales, the precious stones and clay that was enacted by two women (see above page 47). After a break of uncertain length, the text resumes with the passage already quoted above (see page 48-49), where they bring the bones into the "Stone-house" and lay them on a bed.

At this point a seated effigy of the deceased was made, which was from then on to be driven around on a cart between the various locations where the rites were performed. The purpose of the ensuing ritual seems to have been to secure for the deceased an afterlife modelled on his former existence without, of course,

---

89 An emendation to 'shekels' seems to be ruled out because 20 shekels equals $1/2$ mina; cf. H. Otten, ibid. 67 and 142. A smaller weight than the shekel (12.8 gr.) does not seem to be attested, though; cf. Th.P.J. van den Hout, "Masse und Gewichte. Bei den Hethitern," *RlA* 7: 525-527.

90 For funeral meals in general see the interesting remarks by L.-V. Thomas, *Encyclopedia of Religion*, vol. 5, 455.

91 This common but very problematic expression may indicate "drink to the honor of, toast to"; see most recently the bibliographic overview in HW² E 30, where this interpretation is discarded, however.

92 KUB XXX 15+ obv. 1-26 (CTH 450, ed. H. Otten, *HTR* 66-69).

all human deficiencies. This was achieved through a selection of all sorts of objects representing the various aspects of life. From the seventh day on (and maybe earlier) each day was devoted to one of these aspects, such as animal husbandry and its secondary products, agriculture, viticulture, and hunting. The objects were ultimately destroyed, that is, broken into pieces or consumed by fire, thus sharing the fate of the king's body. Fire was apparently thought of as a suitable way of transition to the "other" world, where the king was to spend his further existence.[93] It is very unfortunate that we do not know how the ritual ended, because the description of the activities of the fourteenth day have not come down to us. Was the statue finally transferred to a temple as the further object of the ancestor cult or was it also burnt? In the latter case we would, again, be reminded of the "double burning" practice of Roman imperial times: first the body, then the effigy. This effigy was made of wax so that nothing remained after burning it, which was interpreted as final proof that the emperor did not dwell among the living anymore but had ascended to heaven.[94]

THE HITTITE ROYAL FUNERAL: LOGISTICS AND THE EFFIGY

As to the locations of the ritual, the following can be remarked. First of all, the title of the composition ("When a great loss occurs in Ḫattuša") seems to refer to the death of the king in the capital. This, however, may be understood as "when a vacancy on the throne in Ḫattuša occurs" disregarding where the king may have died. Whatever its exact meaning, a fragment of the second day seems to reckon with the possibility that a king died in some other city than the capital (see above page 50). It certainly mentions the possibility that some king's mausoleum is localized outside the capital, and his remains have to be transported. Further localities mentioned in the ritual are: the Stone-house or mausoleum, a "tent", the palace, a "house" with a portico, and finally the "place where the heads of horses and oxen have been burned", reminding us of the animal skulls found at Osmankayası (see above page 53). Among these the location commonly rendered as "tent" plays an important role. According to the text the seated statue of the deceased is transported from the mausoleum(?) to a tent. During the magic rites in

---

93 A recent example of this can, for instance, be found in Jung Chang's book *Wild Swans. Three Daughters of China* (New York 1991), 82, where she describes the death in 1946 of a certain Mr. Liu, shop owner in north-eastern China: "Monks were brought in to read the Buddhist sutra of 'putting the head down' in the presence of the whole family. Immediately after this, the family members burst out crying. From then on to the day of the burial, on the forty-ninth day after the death, the sound of weeping and wailing was supposed to be heard nonstop from early morning until midnight, accompanied by the constant burning of artificial money for the deceased to use in the other world." For such a role of fire see L.-V. Thomas, *Encyclopedia of Religion*, 5: 457: "Acording to many beliefs, fire is the promise of regeneration and rebirth. (...) Through fire, a superior level of existence can be attained."

94 Cf. E. Bickermann, *ARW* 27 (1929), 13-15.

the tent, the statue stays outside on the cart[95] until it is taken down and brought in for a "great meal." Finally, the statue is put back on the cart again and presumably transported back to the mausoleum. Whether this 'tent' stayed at the same place during all the days of the rituals remains unknown. Only on one occasion is the place of the 'tent' clear, to wit, in a 'house' or building, which at any rate is not the same as the mausoleum but may have been part of the/a palace. A courtyard within that palace complex could be thought of. Whatever its exact nature, it was a movable structure which could be "set up."[96]

The dramatis personae can be roughly divided into two groups. Firstly, the so-called "old/wise women", the *taptara*-women and an anonymous relative of the deceased, who carry out the actual magic and ritual acts. Except for possibly the latter, all protagonists are thus female.[97] The *taptara*-women, who come close to being mourning women, are the only ones to be exclusively attested in funerary context. Secondly, there was a host of cultic personnel, assisting the members of the first group: cup bearers, cooks, "waiters", singers with or without an instrument, the so-called ALAN.ZU$_9$-men, who keep calling out *aḫā*, and the *kita*-men, frequently attested in festivals as crying out or reciting as well, but here explicitly said not to do so. Furthermore, there are the anonymous women gathering the remains of the dead after burning.

---

95 Because of the occasional addition *ašannaš* (lit.) "to sit on" to the sumerogram ^{GIŠ}GIGIR, the translation "cart" is preferred here, although the translation "chariot" cannot be excluded; for the definition of a cart "as an animal-drawn two-wheeler, designed to carry a stable load, i.e. goods or seated passengers" see J. Crouwel, "Carts in Iron Age Cyprus," *Reports of the Department of Antiquities Cyprus* (1985), 203 with lit. Especially interesting are the first millennium representations of persons sitting on but not driving a vehicle in funeral processions such as on the so-called Heroon G in Xanthos (see H. Metzger – P. Coupel, *Fouilles de xanthos. Tome II: L'Acropole Lycienne* (Paris 1963), Pls. XXXIX 1 and XLI 1), and on the south wall of the Karaburun tomb 2 (see M. Mellink, "Excavations at Karataş-Semayük, Lycia 1972," *AJA* 77 (1973), 298-301 and "Excavations at Karataş-Semayük and Elmalı, Lycia 1973," *AJA* 78 (1974), 356-357). Both stem from the early fifth century BC. According to J. Crouwel, "Carts in Iron Age Cyprus" 206 n. 11, these "wheeled thrones" are chariots "with a separate throne placed inside" them. J. Borchardt, *IstMitt.* 18 (1968), 179-186, quoting Herodotus on Scythian burial customs interprets the persons on these wheeled thrones not as the dead portrayed as if alive but as real representations of the deceased, whose body is bound tight on top of the vehicle. In this respect Otten's summary of the Hittite funerary ritual is not always carefully quoted and the alleged striking correpondences between the Hittite text and Herodotus' accounts of Scythian and Thracian customs are much overrated. The possibility, however, that what we see transported is an effigy of the deceased as amply illustrated in the Hittite ritual, must seriously be considered. The same possibility exists for the actual carts and chariots found in the Cypriote Salamis tombs of the eighth and seventh century BC; their use as hearses seems to be rightly questioned by J. Crouwel, ibid. 212-214. The assumption of a Hittite tradition living on in the first millennium is not even necessary, because the use of effigies at funerals seems to be a near universal (see below).

96 For ^{GIŠ}ZA.LAM.GAR "tent" see I. Singer, *StBoT* 27, 100 with literature.

97 On the high number of women in rituals cf. the remarks made by G. Beckman, "From Cradle to Grave: Women's Role in Hittite Medicine and Magic," *JAC* 8 (1993), 36-39.

Apart from these human actors quite a number of animals are mentioned. During the seven days that have fairly completely come down to us, a minimum of ten bovines and 67 sheep are explicitly mentioned as being slaughtered; further, a billy goat, some birds and an unspecified number of horses and mules are burned.

Ingredients used are water, beer, wine, oil (always of "fine" quality), the so-called *ualḫi*-drink, olives, figs, raisins, cheese, meat and innumerable bakery products. Finally, there are several objects, that seem to fulfil a symbolic function, as we will see later on.

One of the most central parts, however, was played by the effigy. A recently recognized text[98] shows how in all likelihood an effigy of the deceased was made on the second day of the ceremonies, that is, immediately after the body had been burned. The fact that the body was not embalmed but practically burned on the day it died may have had its practical and/or religious reasons. Even refined embalming techniques would not have prevented the unpleasant sight of a corpse during fourteen days after a severe illness or a violent death. But the ideology of an immortal king required his presence after the death of his "body natural" for a time long enough to carry out all rites as prescribed and to secure a smooth transition. This function was fulfilled by the effigy that from now on represented the king and was considered living since it received food and drinks and partook in the ceremonies.[99] By transcending, as it were, death, the effigy was at the same time the visual expression of the continuity of kingship and the immortality of the individual monarch. Its use must have posed no problem to the Hittite mind because it was the same principle of substitution that also underlies the substitution ritual we noted earlier. Whether the effigy was in any way made life-like (e.g. by way of a death mask) is unknown, but we do know that it was a seated[100] statue with the sex of the deceased symbolized in the objects it held: a bow and arrows for a man, distaff and spindle for a woman. These objects are well-known in Hittite texts as symbols for masculinity and femininity and immediately recall such passages as in the Paškuwatti ritual against impotence: "I put a spindle and distaff in the patient's hand and he passes under the gate. When

---

98 KBo XXV 184//KUB XXXIX 22, cf. Th.P.J. van den Hout, forthcoming.

99 The procedure of first cremating the body and then replacing it by an effigy as the object of further ceremonies strikingly recalls the already mentioned imperial Roman funerary cult with (at least in some cases) its distinction between a *funus in corpore* and a *funus in effigie*; cf. E. Bickermann, *ARW* 27 (1929), 1-34, and R.E. Giesey, *The Royal Funeral Ceremony* 147-154. For the use and role of funeral effigies in medieval and renaissance England and France see P.S. Fritz, in *Mirrors of Mortality,* ed. J. Whaley, 74-75, R.E. Giesey, *The Royal Funeral Ceremony* 79 sqq., and E.H. Kantorowicz, *The King's Two Bodies. A Study in Mediaeval Political Theology* (Princeton 1962), 419-437; for more general remarks and non-Western examples see L.-V. Thomas, *Encyclopedia of Religion*, 5: 451, 454. Note also the French custom in the second half of the 16th century and the early 17th century to serve the king's body and/or effigy his daily meals: see Giesey, *The Royal Funeral Ceremony* 145-175.

100 For possible seated statues of French Renaissance kings cf. R.E. Giesey, *The Royal Funeral Ceremony* 117-118, but they normally seem to have been portrayed lying.

he steps out of the gate, I take from him spindle and distaff and give him bow [(and) arrows]. Then I speak as follows: 'Behold, I have taken away from you femininity and I have given you back masculinity.'"[101] We are likewise reminded of some of the first millennium neo-Hittite funerary stelae on which we see depicted not only eating and drinking scenes but occasionally men and women with said attributes as well. Although caution is called for in comparing second and first millennium material, W. Orthmann in his study on neo-Hittite sculpture already discussed the possibility of a link with funeral rites in an attempt to explain some of its iconographical features.[102] Now that the symbolism of masculinity and femininity has turned up for the first time in a funerary context in our cuneiform Hittite sources, his suggestion deserves further study.[103]

Interesting in connection with the effigy and the supposed presence of the deceased king is the role of his successor. Among the participants in the ritual listed above, a king (or queen) does not seem to play any role of significance in our preserved material. A king, who according to H. Otten might be the successor to the deceased, is only mentioned in KUB XXXIX 9, possibly describing the events of the fourth day.[104] Unfortunately, however, the highly fragmentary tablet does not allow us to understand what this king is doing and who exactly he is. The only person otherwise coming close to the successor is the *antuḫšaš ŠA* MÁŠ-*ŠU* "the person of his family/clan" occurring once in the 'ritual of cutting the vine' on the twelfth day. This almost complete absence of the new ruler is striking, but may find an explanation in combination with the prolonged life of the former king by way of the effigy. Depending on royal ideology, in some cultures like Renaissance France the simultaneous presence of two kings, one dead but present in the effigy, and the other his successor, was deemed impossible, that is, contradictory to the idea that there can only be a single king embodying perennial kingship. Sometimes, too, it was felt that a king should not be seen mourning or showing any emotions in general.[105] In such cultures it is often customary for the successor to stay almost completely out of sight during the lying-in-state of the dead king and the funeral service.[106] In Hittite practice, too, the deceased king was still present in the seated image, while under 'normal' circumstances the successor would have been known to everybody involved, either named as such

---

101 KUB IX 27+ i 20-27 (CTH 406; ed. H.A. Hoffner, "Paskuwatti's Ritual Against Sexual Impotence," *Aula Orientalis* 5 (1987), 272 and 283 with literature).

102 W. Orthmann, *Untersuchungen zur späthethitischen Kunst* (Bonn 1971), 377-380.

103 For more on this see Th.P.J. van den Hout, forthcoming.

104 Cf. H. Otten, *HTR* 52-55.

105 Cf. J. Varela, *La muerte del rey. El ceremonial funerario de la monarquía española (1500-1885)* (Madrid 1990), 128-132, R.E. Giesey, *The Royal Funeral Ceremony* 46-48. At least he should not be seen mourning his predecessor who was not considered to have died. Hittite kings did occasionally show grief at the death of their next of kin: e.g. Šuppiluliuma I cried when the news of his son's death was broken to him, and Muršili II tells how he himself wept at the demise of his brother Šarrikušuh, the king of Kargamiš.

106 Compare R. Giesey, *The Royal Funeral Ceremony* 41-49, R. Huntington – P. Metcalf, *Celebrations of Death*, 167.

by the old king before dying or even having reigned already for some time next to the latter as *rex designatus*. Maybe here, too, the successor was supposed to keep a low profile as long as the ritual for his predecessor was still being carried out until his final apotheosis.

THE HITTITE ROYAL FUNERAL: THE EIGHTH DAY

Finally, the description of the ritual on the eighth day may serve as an example illustrative of the way the several rites were performed and also of how the different text series supplement each other.

First of all, the outline series states the 'theme' of the day: "[The eighth day: The pi]g divert[s] water [and] they [cut out a piece of meadow]."[107] The restoration in the second part of the 'theme' about the meadow is taken from the ration list where the 'theme' is repeated and the true identity of the 'pig' is revealed:

> On the eighth day the pig diverts water and [they] cut out (a piece of) meadow. A pig's mouth (of) silver, (weighing) ten shekels, a well (of) silver[108], (weighing) twenty shekels, a pickaxe, a shovel at three places inlaid with s[ilver], three conduits inlaid with silver, five cups (of) silver, [...], one of which (with a$^?$) go[ld ...], fourteen pebbles, seven of which [..., seven of which] ... .[109]

Fig 5

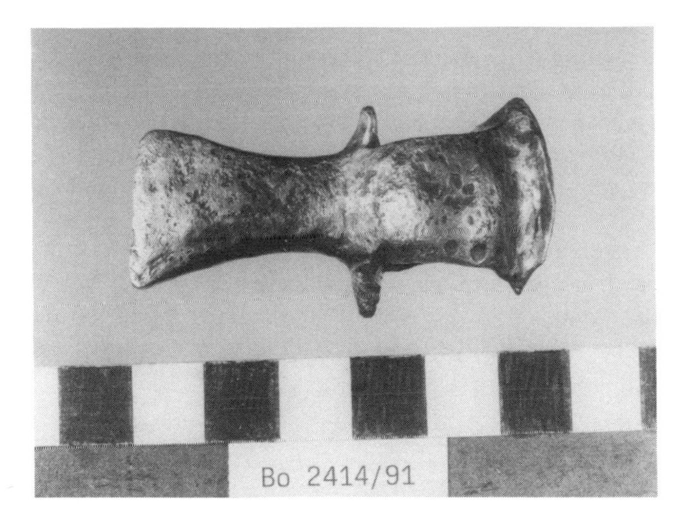

Miniature bronze axes found in the sacrificial pit in the Südbau (Photos courtesy Dr.-Ing. P. Neve).

---

107 KUB XXXIX 6 ii 7'-8' (ed. H. Otten, *HTR* 48-49).

108 A silver well is also attested in KBo XXXIV 66 obv. 3' A-N]A PÚ KÙ.BABBAR-*ma a-wa*[-*an*, recalling KUB XXXIX 35+KUB XXX 24a i 13' (see immediately below).

109 KUB XXXIX 45 obv. 5-9 (cf. H. Otten, *WO* 2 (1954-1959), 477-478).

Fig 6

Miniature bronze axes found in the sacrificial pit in the Südbau (Photos courtesy Dr.-Ing. P. Neve).

Without this description we would at least have visualized the rites being performed at a real well, but it now turns out that the ritual was completely carried out by way of symbolic objects. The "pig's mouth" may have been some tool, called by that name,[110] or a *rhyton*, i.e. a libation vase, in the form of a pig's head. One such pig (or better: boar) rhyton was found at Kültepe, ancient Kaneš/Neša, and roughly dates from the 18th century BC. It is made of clay and measures 6.6 cm in height.[111] Two? pig rhytons of bronze are further mentioned in KUB XLIV 6 obv. 3' and 4'.[112] According to the weight indicated – a pig rhyton of ten shekels equalling ca. 130 gr, and a well of 20 shekels equalling ca. 260 gr – these objects must have been comparatively small: the famous silver stag rhyton in the Schimmel collection weighs ca. 320 gr and is 17 cm long and 18 cm high.[113] So the pig rhyton might have matched in height the one from Kültepe fairly well. The pickaxe, shovel and conduits probably were miniature objects, too, like the two bronze miniature axes (see Figs. 5-6) from Ḫattuša weighing 53 and 50 gr respectively and found in what may have been a sacrificial pit[114] in the "Südbau." The so-called *arda*-birds, figuring in the main ritual description below, may not have been real birds either but, for instance, made of richly inlaid wood as the text seems to indicate. For this we may compare the 35 or 36 *laḫḫanzana*-birds used on the thirteenth day of the ritual.[115] Ten of these were made of wood plated with silver, five of which had extra inlays of gold on their heads, ten were made of wool and another ten of dough; these thirty birds were supplemented by "five or

---

110 Cf. H. Otten, *OLZ* 57 (1962), 231: "(das Gerät) die 'Schweineschnauze' ".

111 See T. Özgüç, *Kültepe-Kaniş. New Researches at the Center of the Assyrian Trade Colonies* (Ankara 1959), 63-64, Taf. 47 Abb. 3-4.

112 In a fragmentary context; cf. H. Ehelolf, "Zu Amarna Knudtzon Nr. 29, 184 und 41, 39ff.," *ZA* 45 (1939), 71-72 ("Bo 490").

113 O.W. Muscarella, *Ancient Art. The Norbert Schimmel Collection* (Mainz 1974), no. 123.

114 See P. Neve, *AA* 1992, 317-319; the interpretation of the pit as sacrificial was put forward by I. Bayburtluoğlu apud Neve, loc.cit. Certainly a miniature axe was the silver one mentioned on the twelfth day of the ritual (XXX 19+ i 55, cf. H. Otten, *HTR* 34-35) and weighing twenty shekels.

115 KUB XXXIX 7 ii 7-14 (cf. H. Otten, *HTR* 36-37).

six" live ones. The "(piece of) meadow" probably was a real piece of earth, because later on it is said to be "poured out" or crumbled.[116] The horses and mules, finally, are only mentioned within the direct speech as confirmed by the liturgy:[117] "When the pig d[iverts] water [the ... says: ' ...] horses (and) mules [... ']"[118] When [they] cut out a (piece of) meadow [the ... says: ' ... on$^{?}$ the] meadow [may] horses (and) mul[es ... '].''[119] We can conclude that the eighth day of the ritual was devoted to animal husbandry and seeks to secure for the deceased the eternal meadows to let his cattle graze on.

Now that the theme, the ingredients, and the words to be spoken are set, the ritual itself can begin.[120] After some initial, very fragmentary lines mentioning among other things the pickaxe and the shovel, the pig rhyton is put into action:[121]

i 5'-10' [The]n, one pig [and] five *arda*-birds [they$^{?}$ ...]. In the silver pig's mouth [they$^{?}$ ...]. Around the *arda*-birds five s[ilver] cups [?] weighing 5 shekels [they$^{?}$] pla[ce. ...$^{?}$] and the *arda*-birds [are placed$^{??}$] around the well. Then the pig diverts water and thus [says/say ...]:

11'-13' 'Behold, the pig has diverted water, and may (it) be [...] to yo[u]. [May$^{?}$] oxen, sheep, [ho]rses, (and) mules satis[fy their thirst$^{?}$[122]].' One ox (and) seven sheep (= obj.)[...] slaug[hter(s)] down into the well.

---

116 E. Masson, *Les Douze Dieux* 35, mentions "l'offrande sous forme de prairie en argent", which finds no support in the text.

117 KUB XXXIX 41 rev. 4'-7' (cf. H. Otten, *OLZ* 57 (1962), 231).

118 Compare below in the main description of the ritual i 12'.

119 Compare below in the main description of the ritual ii 3-4.

120 KUB XXXIX 35+XXX 24+24a+XXXIV 65//KUB XXXIX 36(+)37(+)38(+)39(+)40; with H. Otten, KBo XXXIV Inhaltsübersicht v, nr. 56 may join KUB XXXIX 35+ indirectly, because it has — among other things — the "fourteen stones" in two pairs of seven in common with the ration list. For the possible position of the fragment within the tablet see the join sketch by S. Košak, *Konkordanz der Keilschrifttafeln I. Die Texte der Grabung 1931* (Wiesbaden 1993; = *StBoT* 34), 70.

121 In the following translation each section indicated by line numbers corresponds to a "paragraph" on the Hittite tablet, being each passage that is separated by the scribe from another by a horizontal line drawn on the tablet. The line count differs from the one in H. Otten, *HTR* 58, because of the inclusion of KUB XXXIX 35 at the beginning of the text; line i 10' here = *HTR* 58 line 2'. Italicized are the Hittite (and Akkadian) words that cannot be translated, capitals indicate a Sumerogram without a satisfying translation; square brackets mark breaks in the text on the tablet, round brackets contain words that are supposed to help the understanding of the English text. For another translation of the eighth day see L. Christmann-Franck, *RHA* 29 (1971), 70-72 and G. del Monte in *Archeologia dell'Inferno* (ed. P. Xella), 98-100; the passage i 1'-ii 7 is translated also by E. Masson, *Les Douze Dieux* 34-35.

122 For *šakruwai-* "to water (animals)" see H.G. Güterbock, "Lexicographical Notes," *RHA* XV/60 (1957), 4-6; almost all attestations come from the horse training texts with the animals as object. In contrast to the numerous attestations there the passage in the funerary ritual seems to be the only one with the particle *-za*. L. Christmann-Franck, *RHA* 29 (1971), 71 ("que ... soient a[(breu)vés!"; id. G. del Monte in *Archeologia dell'Inferno* (ed. P. Xella), 99, and E. Masson, *Les Douze Dieux* 34) apparently restores *šakru[uanteš ašandu]*. The particle *-za*, however, seems to plead against this. I therefore propose to restore *-za ... šakru[uandu]* lit. "may they water themselves."

14'-19' Then the Old[woman ...] ... three warm breads, one thick bread (and) cheese and on a p[air of sc]ales [there lies?] (a vase of?) glass?. [... the Ol]d woman holds [...] up towards the Sungod [and speak]s thus: 'Behold, for you we have [...] these [.... Let nobody] depr[ive] him of them, and le[t nobody co]ntest (them against) him.'

20'-22' [Then, o]n top of the cart-to-sit-on to the effigy [they ...], while the singers of the Sungod of Heaven [sing] (accompanied) by a [.... The cup be]arer breaks a sour dough bread and subsequently [...-]s it.

23'-27' [...] and the five *arda*-birds (= obj.) in the sacrificial pit [.... ...] from it? they remove, from the birds (and) cups [the ...] and [the Old] woman remove the si[lver?? (and) gold] and from the conduits they remove [the ...]; they break them into little pieces[123] and [to? ...] the[y] take? them.

The expression *arḫa danzi* "they remove", used at least three times in the preceding paragraph, is interpreted here as indicating the stripping of the birds, cups and conduits (and the other tools?) of their precious inlays before they are destroyed; for *arḫa dā-* in such a context compare the description of the twelfth day, KUB XXX 19+ i 15, 47-50 (cf. HTR 34-35) where a tray? with gold and silver inlays is first stripped, then broken and finally thrown in the hearth. The object "them" at the end must refer to the objects stripped of their gold, silver and precious stones. Like the gold and silver from the tray, it may be that they were brought to the 'Stone-house.'

28'-31' [...] a jug of wine they break and the pickaxe (and) the shovel they [b]urn on that spot. The ashes, however, they pick up and pou[r] them out there, where the horses' heads (and) the oxen's [head]s were burnt.

32'-35' Then they take a pickaxe (and) shovel – those of wood – and cu[t o]ut a (piece of) meadow. The Old woman [lays] the (piece of) meadow on a thick bread [and the ...] hold [them] up towards the [Su]n[god and] the sc[ales] they hold [up to]wards [him?] and th[us ... speak(s)]:

One set of the two pickaxes and shovels must be the one mentioned in the ration list above, but it is not clear which pair. The fact that the first pair is burnt is not a severe objection because they may have been stripped of their inlays already. The place "where the horses' heads (and) the oxen's [head]s were burnt", of course, recalls the find of exactly these animal's heads at the cemetery of Osmankayası (see above page 55).

123 For this meaning of the verb *kinae-* see N. Oettinger, *Die Stammbildung des hethitischen Verbums* (Nürnberg 1979), 162; a different meaning ("assembler, assortir") and therefore different interpretation is followed by E. Masson, *Les Douze Dieux* 40 n. 29, and 35 respectively,

ii 1-7 'This (piece of) meadow, O Sungod, have it made rightly his! Let nobody deprive him of it, let nobody contest it, and may oxen and sheep, horses (and) mules graze on this meadow for him.' They bring the (piece of) meadow there, where the horses' heads (and) the oxen's heads were burnt, and they pour it thereon. The Old woman, in the meantime, takes a thick bread of one *sutu* (weight[124]).

8-11 Then they give the effigy on the cart-to-sit-on to drink, and he$^{?}$ drinks (to) the Sungod. The singer sings accompanied by the *ḫunzinar*-instrument, the ALAN.ZU$_9$-men call out *aḫā* in a whisper, the *kita*-man does not shout.

One wonders whether the subject in the third person singular *ekuzi* "he/she drinks" is the effigy portrayed as if alive (for this see also the next paragraph) or an impersonal subject ("they", German *man*/French *on*) for which one normally would expect a third person plural. A similar singular with a supposedly impersonal meaning occurs below in lines 22, 25, 26, 30 and 31; there, however, it seems less likely that the effigy could be the subject, so that "they" is preferred in the translation.

12-16 They take the effigy down from the cart-to-sit-on. A tent has already been set up on that same spot. They bring it into the tent and seat him on a golden throne, if it (is) a woman, however, they seat her on a golden stool.

The word for "effigy" is written here – as almost everywhere within this ritual – with the Sumerogram ALAM = Hitt. *ešri-*. The latter is known as a word of the neuter gender, which matches the Hittite anaphoric pronoun *-at* in line 14 ("They bring *it* into the tent"). Immediately afterwards, however, it is referred to with the common gender pronoun *-an* in the lines 14 and 16 ("and seat *him* on a golden throne, if it (is) a woman, however, they seat *her*") once again showing how the effigy was considered to be an animate being. The distinction between male and female here regarding the seat occurs at other places in this ritual as well; the stool for a dead queen, i.e. a seat with a very low back or with no back at all, is best illustrated by the small pendant of the seated goddess in the Schimmel collection.[125]

17-22 The main meal of the day is announced and they give the assembled *taptara*-women *šaramma*-bread and a dressed cup bearer gives them each to drink. He drinks (to) the Sungod, the Stormgod, the Tutelary-deity, the Sungoddess of the Earth, each separately, once. The si[ng]er sings (accompanied) by the Ištar-instrument, the ALAN.ZU$_9$-men call out *aḫā* in a whisper (and) they break thick breads.

23-29 Afterwards they drink (to) the grandfathers (and) grandmothers, the sing-

124 The *sutu* is an Akkadian liquid and dry measure of ca. 8.4 ltr.

125 O.W. Muscarella, *Ancient Art*, no. 125.

ers sing (accompanied) by the Ištar-instrument, the ALAN.ZU_9-men call out *aḫā* in a whisper (and) they break thick bread. Afterwards they drink [his] soul three times, the singers [sing] (accompanied) by the Ištar-instrument, the ALAN.ZU_9-men call out *aḫā* in a whisper, the *kita*-man does not shout. Two sweet thick breads they break.

30-34 When they drink his soul for the third time, he says [...], thick bread is not broken. The cup bearer smashes the *išqaruḫ*-vase on the ground, the[n he si]ings? and the assembly [start]s wail[ing].

35-38 They bring the effigy out of the tent, and seat it on the cart-to-s[it-on]. The *tapt*[*ara*]-women fol[low] and [they bring] it [...] and seat it [on ...].

The last paragraph can be partly restored after a similar passage in the description of the twelfth day (KUB XXX 19+ i 61-63, cf. also ibid. i 6-8; ed. H. Otten, *HTR* 32-35). That same passage marks the end of the twelfth day, so that we may assume the same for the eighth day here. This implies that col. iii already describes the events of the ninth day, which cannot be proved because of its very fragmentary state.[126] The only extant fragment of the fourth column of the tablet, KUB XXXIX 35 iv, does at any rate match some of the elements mentioned in the ration list for the ninth day and can therefore with certainty be ascribed to that day.

CONCLUDING REMARKS

In this article we have tried to show that except for some well known but in their variety not very numerous references to royal death implying the apotheosis of the members of the ruling dynasty, quite a few texts suggest that before his ascension the king may have entered the realm of the Netherworld. There is also reason to believe that a certain dichotomy was upheld between the king's bodily remains and his "soul": the body natural and the body politic as reflected in two architectural structures, a mausoleum and a memorial temple. The same distinction may be found in the royal funerary ritual where the king's body was burnt on the first day to be replaced for the remaining thirteen days by an effigy representing the deceased as still dwelling among the living. Although mainly in the footnotes, we took the liberty of pointing several times at similar phenomena in other cultures such as imperial Rome and Renaissance France. Such typological comparisons are useful in the sense that they make us aware of certain problems or hint at a possible significance of otherwise unexplained elements in the Hittite ritual.

---

126 One might point at the (legal) formula "Let nobody deprive him of it, let nobody contest it" occurring iii 5'-6', which is found relatively early in the description of the eighth day also (i 18'-19', but also ii 2-3) when the objects symbolizing the theme of the day are introduced and held up to the Sungod. In the case of the ninth day one is tempted to read the GA in iii 3' immediately preceding this formula as the sumerogram for "milk", which seems to have been the main item in the theme of that day according to the ration list (cf. H. Otten, *WO* 2 (1954-1959), 478-479).

At the outset of this contribution it was said that for the beliefs of the 'average' Hittite we seem to lack almost completely written information. Fortunately, however, archaeology is able to fill that gap in our knowledge to a certain extent. There may be one text, although of possible Mesopotamian origin, that draws a sad picture of a stay in the Netherworld:

One doesn't recognize the other. Sisters by the same mother do [not re]cognize each other. Brothers by the same father do [not re]cognize each other. A mother does [not] recognize [her] own child. [A child] does [not] recognize [its own] mother. ... From a fi[ne] table they do [no]t eat. From a [fi]ne stool they do [n]ot eat. From a [f]ine cup they do not drink. They do not eat [goo]d food. They do not drink my? good drink. They eat bits of mud. They [dri]nk muddy waters(?).[127]

Seen in this light the anxiety of the king facing the dark beyond death is not surprising. But for him the stay may have been of only a very short duration. He had the privilege of royal death.

127 KBo XXII 178(+)KUB XLVIII 109 ii 4'-iii 7'; translation H.A. Hoffner, *A Scientific Humanist. Studies in Honor of Abraham Sachs*, ed. E. Leichty – M. de Jong Ellis (Philadelphia 1988), 192.

BIBLIOGRAPHY AND ABBREVIATIONS

AA — *Archäologische Anzeiger*

ABoT — *Ankara Arkeoloji Müzesinde bulunan Boğazköy Tabletleri* (Istanbul 1948)

AfO — *Archiv für Orientforschung*

AION — *Annali dell'Istituto Universitario Orientale di Napoli*

AJA — *American Journal of Archaeology*

AoF — *Altorientalische Forschungen*

Archeologia dell'Inferno — *Archeologia dell'Inferno. L'Aldilà nel mondo antico vicino-orientale e classico*, ed. P. Xella (Verona 1987),

Archi, A. — "Divinités Sémitiques et Divinités de Substrat. Le Cas d'Išḫara et d'Ištar à Ebla", *MARI* 7 (1993), 71-78

ARW — *Archiv für Religionswissenschaft*

AuOr — *Aula Orientalis*

G. Beckman, *StBoT* 29 — *Hittite Birth Rituals* (Wiesbaden 1983)

—, — "From Cradle to Grave: Women's Role in Hittite Medicine and Magic," *JAC* 8 (1993), 25-39

Bh. — Beiheft

J. Bickerman, — "Die römische Kaiserapotheose", *ARW* 27 (1929), 1-34

BiOr — *Bibliotheca Orientalis*

K. Bittel et al., — *Die hethitischen Grabfunde von Osmankayası* (Berlin 1958)

—, — *Die Hethiter* (München 1976)

J. Borchardt, — "Epichorische, gräko-persisch beeinflußte Reliefs in Kilikien" *IstMitt.* 18 (1968), 161-211

J. Catsanicos, — *Recherches sur le Vocabulaire de la Faute. Apports du Hittite à l'étude de la phraséologie indo-européenne* (Cahiers de N.A.B.U. 2, Paris 1991)

CHD — H.G. Güterbock – H.A. Hoffner, *The Hittite Dictionary of the Oriental Institute of the University of Chicago, L-N* (Chicago 1989)

L. Christmann-Franck, — "Le Rituel des Funérailles Royales Hittites", *RHA* 29 (1971), 61-111

J. Crouwel, — "Carts in Iron Age Cyprus" *Report of the Department of Antiquities Cyprus 1985* (Nicosia, Cyprus) 203-221 with Plates XXXI-XXXIV.

CTH — E. Laroche, *Catalogue des Textes Hittites* (Paris 1971)

C.-M. Edsman, — "Fire" in *Encyclopedia of Religion*, vol.5, 340-346.

H. Ehelolf, — "Zu Amarna Knudtzon Nr. 29, 184 und 41, 39ff.", *ZA* 45 (1939), 70-73

M. Eliade, — see *Encyclopedia of Religion*

Encyclopedia of Religion — *Encyclopedia of Religion*, ed. in chief M. Eliade (New York-London 1987)

J. Friedrich, SV — *Staatsverträge des Hatti-Reiches in hethitischer Sprache* (Leipzig 1926)

P.S. Fritz, — "From 'Public' to 'Private': The Royal Funerals in England, 1500-1830" in *Mirrors of Mortality. Studies in the Social History of Death*, ed. J. Whaley (London 1981), 61-80

FsÖzgüç — *Anatolia and the Ancient Near East. Studies in Honor of Tahsin Özgüç*, ed. K. Emre – B. Hrouda – M. Mellink – N. Ögüç (Ankara 1989)

FsPolomé — *Perspectives on Indo-European Language, Culture and Religion. Studies in Honor of Edgar C. Polomé* (McLean, Virginia 1989)

FsPope — *Love and Death in the Ancient Near East. Essays in Honor of Marvin Pope*, ed. J.H. Marks – R.M. Good (Gilford, Connecticut 1987)

R.E. Giesey, — *The Royal Funeral Ceremony in Renaissance France* (Genève 1960)

A. Götze, *Ḫatt.* — *Ḫattušiliš. Der Bericht über seine Thronbesteigung nebst den Paralleltexten* (Leipzig 1925)

—, *AM* — *Die Annalen des Muršiliš* (Leipzig 1933)

A. Goetze, — review of H. Otten, *HTR* in *AJA* 64 (1960) 377-378

H.G. Güterbock, — "The Song of Ullikummi. Revised Text of the Hittite Version of a Hurrian Myth," *JCS* 5 (1951), 135-161, ibid. 6 (1952), 8-42

—, "Lexicographical Notes", *RHA* XV/60 (1957), 1-6
—, "The Hittite Conquest of Cyprus Reconsidered", *JNES* 26 (1967), 73-81
—, see also CHD
O.R. Gurney, *Schweich* *Some Aspects of Hittite Religion* (The Schweich Lectures 1976, Oxford 1977)
—, *Hittites* *The Hittites* (Harmondsworth 1990)
Haas, V. – Wäfler, M., "Bemerkungen zu [É]*hešti/ā-*", *UF* 8 (1976), 65-99, *UF* 9 (1977), 87-122
A. Hagenbuchner, *THeth.* 16 *Die Korrespondenz der Hethiter, 2. Teil. Die Briefe mit Transkription, Übersetzung und Kommentar* (Heidelberg 1989)
J.D. Hawkins, "Late Hittite Funerary Monuments", *Mesopotamia* 8 (1980), 213-225
H.A. Hoffner, "Paskuwatti's Ritual Against Sexual Impotence (CTH 406)", *AuOr* 5 (1987), 271-287
—, *Hittite Myths* "Hittite Terms for the Life Span", *FsPope* 53-55
—, "A Scene in the Realm of the Dead", *MemSachs* 191-199
—, *Hittite Myths, translated by H.A. Hoffner, Jr.* (Atlanta, Georgia 1990)
—, see also CHD
Th.P.J. van den Hout, *Tudḫalija Kosmokrator. Gedachten over ikonografie en ideologie van een Hettitische koning*. (Amsterdam 1993)
—, *StBoT* 38 *Der Ulmitešub-Vertrag. Eine prosopographische Untersuchung* (Wiesbaden 1994)
—, forthcoming "An image of the Dead?", in , Proceedings of the II Congresso Internazionale di Hittitologia (Pavia 1992), ed. O. Carruba – C. Mora
Ph.H.J. Houwink ten Cate, "The Bronze Tablet of Tudhaliyas IV and its Geographical and Historical Relations", *ZA* 82 (1992), 233-270
HS *Historische Sprachforschung*
R. Huntington – P. Metcalf, *Celebrations of Death. The Anthropology of mortuary Ritual* (Cambridge 1979)
M. Hutter, *Behexung, Entsühnung und Heilung. Das Ritual der Tunnawiya für ein Königspaar aus mittelhethitischer Zeit (KBo XXI 1 – KUB IX 34 – KBo XXI 6)* (Freiburg-Göttingen 1988)
—, "Bemerkungen zur Verwendung magischer Rituale in mittelhethitischer Zeit", *AoF* 18 (1991), 32-43
HW² J. Friedrich – A. Kammenhuber, *Hethitisches Wörterbuch*, 2. Auflage (Heidelberg 1975 ff.)
F. Imparati, "Le Istituzione Cultuali del [NA4]*ḫékur* e il Potere Centrale Ittita", *SMEA* 18 (1977), 19-64
IstMitt. *Istanbuler Mitteilungen*
JAC *Journal of Ancient Civilizations*
JCS *Journal of Cuneiform Studies*
JEOL *Jaarbericht Ex Oriente Lux*
JNES *Journal of Near Eastern Studies*
A. Kammenhuber, "Die hethitischen Vorstellungen von Seele und Leib, Herz und Leibesinnerem, Kopf und Person (I. Teil)", *ZA* 56 (1964), 150-212
—, see also HW²
E.H. Kantorowicz, *The King's Two Bodies. A Study in Mediaeval Political Theology* (Princeton, NJ 1966²)
KBo Keilschrifttexte aus Boghazköy
G. Kellerman, *Recherche sur les Rituels de Fondation Hittites* (diss., Paris 1980)
S. Košak, *StBoT* 34 *Konkordanz der Keilschrifttafeln I. Die Texte der Grabung 1931* (Wiesbaden 1992)
KUB Keilschrifturkunden aus Boghazköy
H.M. Kümmel, *StBoT* 3 *Ersatzrituale für den hethitischen König* (Wiesbaden 1967)
J. de Kuyper, Les Obligations Funéraires d'une Reine Hittite, SEL 4 (1987), 93-99
KZ *Kuhns Zeitschrift/Zeitschrift für vergleichende Sprachforschung*

E. Laroche,	review of H. Otten, *HTR* in *BiOr* 18 (1961), 82-84
—,	"Nişantaş", *Anatolica* 3 (1969-1970), 93-98
G. McMahon, *State Cult*	*The Hittite State Cult of the Tutelary Deities* (Chicago 1991)
M. Marazzi (ed.),	*L'Anatolia Hittita. Repertori archeologici ed epigrafici* (Roma 1986)
MARI	*Mari Annales de Recherches Interdisciplinaires*
St. de Martino,	"Hattušili e Haštayar: un Problema Aperto", *OA* 28 (1989), 1-24
E. Masson, *Les Douze Dieux*	*Les Douze Dieux de l'Immortalité* (Paris 1989)
MDOG	*Mitteilungen der Deutschen Orientgesellschaft*
Melchert, H.C., *Ablative and Instrumental*	*Ablative and Instrumental in Hittite* (diss., Harvard University, Cambridge, Massachusetts 1977)
—,	"Death and the Hittite King", *FsPolomé* 182-188
M. Mellink,	*A Hittite Cemetery at Gordion* (Philadelphia 1956)
—,	"Excavations at Karataş-Semayük, Lycia 1972," *AJA* 77 (1973), 293-307
—,	"Excavations at Karataş-Semayük and Elmalı, Lycia 1973," *AJA* 78 (1974), 351-359
MemSachs	E. Leichty – M. de Jong Ellis (edd.), *A Scientific Humanist. Studies in Honor of Abraham Sachs* (Philadelphia 1988)
P. Meriggi,	"Spigolando nei Testi Storici Etei, Orientis Antiqui Collectio 13 (1978), 67-70
H. Metzger – P. Coupel,	*Fouilles de Xanthos. Tome II: L'Acropole Lycienne* (Paris 1963)
G. del Monte,	"Il terrore dei morti," *Annali dell'Istituto Universitario Orientale di Napoli* 33 (1973), 373-385
—,	—, "La fame dei morti," *AION* 35 (1975), 319-346
—,	"Inferno e Paradiso nel Mondo Hittita" in *Archeologia dell'Inferno. L'Aldilà nel mondo vicino-orientale e classico,* ed. P. Xella (Verona 1987) 95-115
O. Muscarella, *Ancient Art*	*Ancient Art, The Norbert Schimmel Collection* (Mainz 1974)
P. Neve,	"Einige Bemerkungen zu der Kammer B in Yazılıkaya", *FsÖzgüç* 345-355
—,	"Die Ausgrabungen in Boğazköy-Ḫattuša 1991", *AA* 1992, 307-337
—, *Ḫattuša*	*Ḫattuša – Stadt der Götter und Tempel. Neue Ausgrabungen in der Haupstadt der Hethiter* (Antike Welt, Sondernummer 1992)
OA	*Oriens Antiquus*
N. Oettinger, *Stammbildung*	*Die Stammbildung des hethitischen Verbums* (Nürnberg 1979)
—,	"Die *n*-Stämme des Hethitischen und ihre indogermanischen Ausgangspunkte," *KZ* 94 (1980), 44-63
T. Özgüç, *Bestattungs-braeuche*	*Bestattungsbraeuche im vorgeschichtlichen Anatolien* (Ankara 1948)
—, *Cemetery at Kazankaya*	*Excavations at Maşat Höyük and Investigations in its Vicinity* (Ankara 1978)
—, *Kültepe-Kaniş*	*Kültepe-Kaniş. New Researches at the Center of the Assyrian Trade Colonies* (Ankara 1959)
OLZ	*Orientalistische Literaturzeitung*
Or.	*Orientalia*
W. Orthmann,	*Das Gräberfeld bei Ilıca* (Wiesbaden 1967)
—, Untersuchungen	*Untersuchungen zur späthethitischen Kunst* (Bonn 1971)
Otten, H.,	"Die hethitischen 'Königslisten' und die altorientalische Chronologie", *MDOG* 83 (1951), 47-71
—, *HTR*	*Hethitische Totenrituale* (Berlin 1958)
—,	"Eine Lieferungsliste zum Totenritual der hethitischen Könige", *WO* 2 (1954-1959), 477-479
—,	"Eine Beschwörung der Unterirdischen aus Boğazköy", *ZA* 54 (1961), 114-157

—,	"Zu den hethitischen Totenritualen", *OLZ* (1962), 229-233
—,*Quellen*	*Die hethitischen historischen Quellen und die altorientalische Chronologie* (Mainz-Wiesbaden 1968)
—, *StBoT* Bh. 1	Die Bronzetafel aus Boğazköy. Ein Staatsvertrag Tutḫalijas IV. (Wiesbaden 1988)
H. Otten, – J. Siegelová	"Die hethitischen Gulš-Gottheiten und die Erschaffung der Menschen", *AfO* 23 (1970), 32-38
J. Puhvel,	"Hittite *annaš šiwaz*", *KZ* 83 (1969), 59-63
—, *HED*	*Hittite Etymological Dictionary* (Berlin-New York 1984 ff.)
—,	"Shaft-shedding Artemis and mind-voiding Ate: Hittite determinants of Greek etyma", *HS* 105 (1992) 4-8
RHA	*Revue Hittite et Asianique*
RlA	*Reallexikon der Assyriologie und Vorderasiatischen Archäologie*, ed. D.O. Edzard et al. (Berlin-New York)
SEL	*Studi Epigrafici e Linguistici*
I. Singer, *StBoT* 27	*The Hittite KI.LAM Festival, Part One* (Wiesbaden 1983)
SMEA	*Studi Micenei ed Egeo-Anatolici*
StBoT	*Studien zu den Boğazköy-Texten*
G. Steiner	"Die Unterweltsbeschwörung des Odysseus im Lichte hethitischer Texte", *UF* 1 (1969), 265-283
D. Sürenhagen	"Zwei Gebete Ḫattušilis und der Puduḫepa," *AoF* 8 (1981), 83-168
L.-V. Thomas	"Funeral Rites" in M. Eliade (ed.), *Encyclopedia of Religion*, vol. 5.
A. Ünal, THeth. 6	*Ein Orakeltext über die Intrigen am hethitischen Hof (KUB XXII 70 = Bo 2011)* (Heidelberg 1978)
UF	*Ugarit Forschungen*
J. Varela,	*La Muerte del Rey. El ceremonial funerario de la monarquía española* (1500-1885) (Madrid 1990)
WbMyth	H.W. Haussig (ed.), *Wörterbuch der Mythologie* (Stuttgart 1965)
R. Werner,	review of H. Otten, *HTR* in *Or.* 34 (1965) 379-381
—, *StBoT* 4	*Hethitische Gerichtsprotokollen* (Wiesbaden 1967)
WO	*Welt des Orients*
J. Whaley (ed.),	*Mirrors of Mortality. Studies in the Social History of Death* (London 1981)
P. Xella (ed.),	*Archeologia dell'Inferno. L'Aldilà nel mondo vicino-orientale e classico* (Verona 1987)
Yaz[2]	K. Bittel et al., *Das hethitische Felsheiligtum Yazılıkaya* (Berlin 1975)
ZA	*Zeitschrift für Assyriologie*

Death and the After-life in the Hebrew Bible of Ancient Israel

Nico van Uchelen

INTRODUCTION

The Hebrew Bible came into existence centuries after ancient Israel had ceased to exist. During a long process of internal and external growth, the literary corpus went through at least three important historical phases. The three separate periods have left their mark on the character of the book.

The long years of the Babylonian captivity in the 6th century before the Common Era can be assumed to have been an important period of cultural-religious restoration. The gathering, ordering and rewriting of the old-time traditions and the addition of new writings may have produced the framework of what nowadays is called the Hebrew Bible. This harmonizing framework must have pertained not so much to material extent and number of books as to literary character and religious tenure[1].

The period of the Roman occupation and the destruction of the temple, and of Jerusalem, in 70 CE, had their accelerating influence on the final form of the literary material, which resulted in a gradual canonisation of the list of books[2]. Finally, synagogical urgency and philological accuracy induced rabbinic scholars in the 8th century CE to provide the meanwhile well-tried and trusty text with the final signs and signals for pronouncing and reciting.

This procedure has led to a curious paradox. On the one hand the Hebrew Bible presents itself as the "national" history of ancient Israel. The "faits et gestes" of the people of the Lord have been reported chronologically, beginning with the Creation and up to and including the return to the country after the Babylonian captivity. On the other hand, however, the canonical corpus, finally being completed by later periodically redactional activities has become an arte-

---

1 "Difficult as is the task of tracing the growth of the Old Testament literature and disentangling the strands of the several traditions which preceded the written records, that of reconstructing the processes by which the Old Testament Canon emerged is still more complex", cf.G.W.Anderson, "Canonical and non-canonical", *The Cambridge History of the Bible*, Vol.I (*From the Beginnings to Jerome*), Cambridge 1970, Part III. The Old Testament, 113-158.

2 "Denn die beiden entscheidenden Phasen in der Entstehungsgeschichte des Kanons, das babylonische Exil und die Zerstörung des zweiten Tempels, bedeuten nicht nur einen Verlust der Rechtshoheit und politischen Identität, sondern auch der rituellen Kontinuität. Beide mußten in der Form des Kanons gerettet werden, um den Bruch zu überdauern", cf. Jan Astmann, *Das kulturelle Gedächtnis. Schrift, Erinnerung und politische Identität in frühen Hochkulturen*, München 1992, Zweites Kapitel, Schriftkultur, II.Kanon – zur Klärung eines Begriffes, 106.

fact, without a clear literary historical context. The harmonizing procedure has made it impossible to discern a continuous development of religious thoughts and to describe a coherent evolution of ethical ideas. After all, the redactional interventions and proceedings have resulted in a kaleidoscopic document, mainly of a narrative character, containing high theological norms and brought to the fore in a skillful literary context.

In reflection on death and after-life in ancient Israel, one has to be keenly aware of this theological and literary state of affairs before one tries to sketch a historical continuity of biblical ideas and perceptions[3].

Ancient Israel, as the object of 'historical' description in the Hebrew Bible, geographically formed only a tiny part of the "Fertile Crescent". This imagined half-circled area extended from and along the rivers Euphrates and Tigris, encompassing "Mesopotamia" in the East, stretching along the Mediterranean coast in the North-West, via Phoenicia and Kanaän, to the land of Egypt, along the river Nile to the South. As a narrow strip, "the promised land" was thus lying clinched in between mighty empires of high cultural prestige and military power. From time to time the Egyptians, the Babylonians and Assyrians demonstrated their self-assertive prestige and power. Curiously enough, however, the after-effect of the Egyptian and/or Babylonian expansive appetites have left neither cultural nor military footmarks whatsoever in the tiny strip of land west of the river Jordan. From an archeological point of view no monumental remnants from the Egyptian or Babylonian culture have been uncovered up till now. Neither architectural structures of buildings or stylistic modelling of statues, nor syncretistic cultic objects reveal any substantial influence of the powerful neighbours. Only hosts of Egyptian scarabs and a few tablets written in cuneiform Mesopotamian script betray a certain degree of cultural interaction.

It is most curious to observe the fact that any influence in the field of thought and religion is also lacking. This also holds true regarding ideas on death and the after-life. In Egypt as well as in Mesopotamia, concern with death and the after-life was an integral part of daily life, in a material and in a spiritual respect. In both cultures the experience of vulnerable life and relentless death was greatly determined by the natural rhythm of life and death of the vegetation. In Egypt this experience resulted in a dramatic cult-ceremonial[4], whereas the heroic epic of Gilgamesh bears witness to tragic feelings through the ages regarding the way of life and death in Mesopotamia[5].

---

3 For a more religious-historical context, cf. J.N.Tromp, *Primitive Conceptions of Death and the Netherworld in the Old Testament* (Biblica Orientalia 21), Rome 1969.

4 Cf. J. Zandee, *Death as an enemy. Studies in the History of Religion (Supplements to Numen V)*, Leiden 1960; C.J.Bleeker, *Het oord van stilte. Dood en eeuwigheid naar oud-Egyptische geloofsbeleving*, Katwijk 1979.

5 E.Ebeling, *Tod und Leben nach den Vorstellungen der Babylonier*, Berlin 1931; Thorkild Jacobsen, *The Treasures of Darkness.* History of Mesopotamian Religion, Yale Univ. Press, New Haven and London, 1979^{3}, 224.

Not the slightest trace of the Egyptian ritual drama or of the Mesopotamian tragic heroism can be found in the Hebrew Bible of ancient Israel. The pattern of religious feelings in the two extreme areas of the fertile crescent, Egypt in Africa and Mesopotamia in Asia, to a great extent has been stamped and marked by the unrelenting course of nature. The endless change of seasons, that is, the inevitable dying and blooming of the natural vegetation was felt and experienced as the compelling course of divine as well as of human life.

From the first page of the Hebrew Bible, however, "nature" as such, that is, as a divine power, has been desecrated, and the "divinities" as such, that is, the elementary forces of nature, have been demythologized. From the beginning "nature" has been qualified as mere creation. Far from being divine powers in themselves, the dying and blooming of the vegetation, the rising up and falling down of the waters are believed to take place according to the will and the word of the God of Israel. The experience of life and death, the striving for life and the doom of death, has been set free from the squeezing bonds of nature.

The following sketch of death and after-life in the Hebrew Bible has the character of a line pattern. From the state of affairs regarding the character of the Hebrew Bible and therefore from the point of method, a historical line of development cannot be drawn. Regarding the perception of death, three main lines may be distinguished within the sketchy interplay :

1. a theological-anthropological line;
2. a literary metaphorical line;
3. a personal-emotional line.

DEATH

1 The theological-anthropological line

In the meticulous story of what is generally known as "paradise", in Hebrew designated as "the garden of Eden" (Gen.2:8)[6], a sonorous resonance of the precariousness of human life can be heard.

It is true that "the earth brought forth vegetation" (1:12), the waters swarmed with living creatures, trees of every kind bore fruit, birds increased on the earth. And what is more, from the rib of man God fashioned a woman as a fitting helper. But, dangerously enough, "the Lord God formed man (*'adam*) from the dust (*'afar*) of the earth (*'adama*)" (2:7). The risky moment lies in the almost casual word *'afar*.

Here is, so to speak, a snake in the grass. In the Hebrew Bible *'afar* functions as the including denotation of what sticks together like grains of sand, what in the dry season is blown up in dusty clouds, what after burning has become like powder, what has remained after thorough devastation.

---

6 All quotations have been taken from *TANAKH. The Holy Scriptures.*The New JPS Translation According to the Traditional Hebrew Text. The Jewish Publication Society, Phildelphia-New York 1988.

'Afar as the first qualification that comes of man has a portentous sound, notwithstanding the promising statement: "He blew into his nostrils the breath of life, and man became a living being"(2:7). For, after man and woman had given audience to the whisper of the shrewd snake, the story immediately continues:

"By the sweat of your brow
Shall you get bread to eat,
Until you return to the ground (*'adama*) -
for from it you were taken.
For dust (*'afar*) you are,
and to dust (*'afar*) you shall return" (3:19).

Another man in the Hebrew Bible, Job, apparently fears the same destiny. Hoping against hope, he starts a dispute with his God, arguing:

"Considered that you fashioned me like clay,
Will you then turn me back into dust (*'afar*) ?" (Job 10:9).

In an inner-biblical dialogue, it would seem, Kohelet, son of David, king of Jerusalem, gives a statement of reply.

> "So I decided, as regards men, to disassociate them (from) the divine beings and to face the fact that they are beasts. For in respect of the fate of man and the fate of beast, they have one and the same fate: as the one dies so dies the other, and both have the same life-breath; man has no superiority over beast, since both amount to nothing. Both go to the same place; both came from dust and both return to dust (*'afar*). Who knows if a man's life-breath does rise upward and if a beast's does sink down into the earth?" (Eccles. 3:18-22).

It is this deeply rooted feeling that made the Hebrew Bible speak of dying men as people "descending into the dust" (Ps.22:30; Job 17:16) and of dying as "to lie down in the dust" (Job 7:21; 20:11; 21:26). The dead are circumscribed as people "who dwell in the dust" (Is.26:19) or as "those that sleep in the dust" (Dan.12:2). The term "dust" may be seen as the materialization of the underworld. The Hebrew for "underworld" (*she'ol*), the place into which one descends and dwells in, contains a double meaning, on the one hand "to be noisy" and on the other, "to be empty". However, what is dominant in this respect is the geographical denotation: the underworld is a place which is beneath the earth and opposite the sky, has its own depth, accessible through gates and passable by roads.

It is also a place that, greedy and insatiable, is in expectation of newcomers (Is.14:9; Prov. 30:16). Sometimes it is seen as a prison (II Sam.22:6 = Ps.18:6), from which ransom may be possible (Hos.13:14; Ps.49:6). From a theological point of view, the underworld is the very place in which the Lord cannot be praised (Is.38:18; Ps.6:6; 88:4).

This concrete, empirical background gives deep colour to another qualification of man, being only "flesh" (*basar*). This qualification covers two aspects of human existence. On the one hand "flesh" serves to denote man in his bodily appearance and pertains to his visible figure. On the other hand, however, the denotation qualifies him in his weak existence, in his powerlessness and in his mortality. As it has been phrased in Genesis and in Psalms:

> "The Lord said, My breath shall not abide in man forever,
> since he too is flesh" (Gen.6:3).
> "For He remembered that they were but flesh,
> a passing breath that does not return" (Ps. 78:39).

The qualifications "man of earth" (*'adam*) and "man of flesh" (*basar*) may not have stressed human weakness adequately enough, for another designation "man of fragility" (*'enosh*) frequently occurs. Mainly used in poetical texts (Psalms and Job), this designation serves to qualify man in his dependence, in his limitation and in his mortality. Throughout the Hebrew Bible, from the first page up to and including the last, man is conceived of as dependent on the breath of the Lord, as limited in the days of his life and as mortal as the grass, the flowers and the animals.

It may be said that the theological-anthropological point of view does not allow one to design a clear system or a balanced scheme. It is only in a sketchy line that the theological view can be drawn. In the Hebrew Bible mortality and death are not dealt with in paragraphs and articles, but in stories and poems. The vulnerable and finite aspect of human life is seen in the light of the belief in and the conception of God, as conceived by ancient Israel. The main theme of this aspect of human life is realistic: man has to live in an evident vulnerability, finally resulting in the inevitable transitoriness.

2 The literary-metaphorical line

Following the theological-anthropological line the investigator has to be aware of the kaleidoscopic picture the Hebrew Bible offers. However, the changing figures match in one cardinal point of view. God gives life and death to man. Man has been taken from earth and will return to earth.

Following the literary-metaphorical line, the investigator finds himself in a world of words and images. This particular literary state of affairs seriously complicates a cohesive representation of the textual data. It is advisable therefore to look up, to read through and to go after "the turn of the metaphor". In this field the Hebrew Bible at least draws two lines: one concerning the image of man destined to die and the other concerning the image of death as perceived by man.

a. The image of man destined to die particularly occurs in Psalms and Job. From the large and varied amount of texts, two metaphorical aspects will be dealt with. Both give an imaginative description of man inevitably doomed to die. Following the phrasing of the words and figures, the images of 'man as grass' and 'man as breath ' will be presented by quotation.

The image of grass:
Psalm 90:3-6

> 3. You return man to dust
> You decreed, "Return you mortals!'

4. For in Your sight a thousand years
are like yesterday that has passed,
like a watch of the night.
5. You engulf men in sleep;
at daybreak they are grass that renews itself;
at daybreak it flourishes anew;
by dusk it withers and dries up.

Psalm 103:14-16

14. For He knows how we are formed;
He is mindful that we are dust.
15. Man, his days are like those of grass;
he blooms like a flower of the field;
16. a wind passes by and it is no more,
its own place no longer knows it

Job 14:1-2

1. Man born of woman is short-lived and sated with trouble,
2. He blossoms like a flower and withers;
he vanishes like a shadow and does not endure.

Metaphorically, man is seen as grass which withers in the evening and which shrivels up in the wind, not holding out, and the place of which is no longer known.

The image of breath:
Psalm 39:6,7,12

6. You have made my life just handbreadths long;
its span is nothing in Your sight
no man endures any longer than a breath.
7. Man walks about as a mere shadow;
mere futility is his hustle and bustle,
amassing and not knowing who will gather in.
12. You chastise a man in punishment for his sin,
consuming like a moth what he treasures,
No man is more than a breath.

Psalm 62:10

10. Men are mere breath
mortals, illusion;
placed on a scale together,
they weigh even less than a breath.

Psalm 144:3,4

3. O Lord, what is man that you should care about him,
mortal man, that You should think of him?
4. Man is like a breath;
his days are like a passing shadow.

In comparison, man is seen as breath, in Hebrew *hevel*, which means ‘empty air’ or ‘void wind’, that passes on as shade and becomes lost like a moth.

b. The image of death as perceived by man in the Hebrew Bible is often related to a certain quality of life or to several periods of life. As soon as life loses quality, death wins ground. To the perception of ancient Israel, death threateningly manifests itself in various forms during life. Dangerous cases and ominous domains may be full of death in human life. The danger of death displays itself in hours of hostility, in places of prison and in situations of disease. Important expressive terms to characterize these awful dangers of death are 'mighty waters' and 'roaring lions'. Some quotations may illustrate the expressiveness of the texts.

The image of the mighty waters:
Psalm 18:5,17

5. Ropes of Death encompassed me;
torrents of Belial* terrified me;
ropes of Sheol encircled me;
snares of Death confronted me.
17. He reached down from on high, He took me
He drew me out of the mighty waters;
He saved me from my fierce enemy,
from foes too strong for me.

*i.e. the netherworld, like "Death"and "Sheol".

Psalm 88:4-7

4. For I am sated with misfortune;
I am at the brink of Sheol.
5. I am numbered with those who go down to the Pit;
I am a helpless man
6. abandoned among the dead,
like bodies lying in the grave
of whom You are mindful no more,
and who are cut off from Your care.
7. You have put me at the bottom of the Pit,
in the darkest places, in the depths.

The image of the roaring lions:
Psalm 22:13,14,17,21,22

13. Many bulls surround me,
mighty ones of Bashan encircle me.
14. They open their mouths at me
like tearing, roaring lions.
17. Dogs surround me
a pack of evil ones closes in on me,
like lions they maul my hands and feet.
21. Save my life from the sword,
my precious life from the clutches of a dog.
22. Deliver me from a lion's mouth;
from the horns of wild oxen rescue me.

Both groups of texts, concerning 'the mighty waters' as well as 'the wild animals', paint a lively picture of the consciousness that "the gates of death" and "the gates of darkness" (Job 38:17) are set up and opened in the midst of life.

3 The personal-emotional line

The predominantly narrative character of the Hebrew Bible, connected with the often poetical diction, does not always allow us to draw a systematic line. In the narration of the Hebrew Bible people are not portrayed as individuals with personal character-drawing. Abraham, Isaac, Jacob, Moses, Aaron, Gideon, Samson, David and Solomon, instead of being men of flesh and blood, are sketched rather as types, as exemplary figures. Acting as models of belief or of disbelief they do not come to the fore as persons, and they do not play an explicitly historical role.

Sometimes, however, the biblical art of narration gives a glimpse of personal character or of stirred emotions. Faced with misfortune or death, biblical figures are sometimes depicted in their moments of weakness or distress. Two examples may suffice to illustrate the presence of personal-emotional responses in dreadful situations.

The book of Genesis contains the story of Jacob carrying in his hands the blood-drenched garment of his beloved son Joseph. It was brought to him by his other sons with the suggestion that Joseph had been torn up by a wild animal. The suggestion served to hide their jealous wrong-doing by selling Joseph to Egypt. Jacob recognized the garment, saying :

> "'My son's tunic! A savage beast devoured him! Joseph was torn by a beast!' Jacob rent his clothes, put sackcloth on his loins, and observed mourning for his son many days. All his sons and daughters sought to comfort him; but he refused to be comforted, saying, 'No, I will go down mourning to my son in Sheol.' Thus his father bewailed him"(Gen. 37:33-35).

Finally, an episode from the life of David, taken from the book of Samuel. He has received the tidings that his bosom friend Jonathan has been killed in a battle with the Philistines:

> "David took hold of his clothes and rent them, and so did all the men with him. They lamented and wept, and they fasted until evening for Saul and his son Jonathan"(2 Sam.1:11,12).
>
> "And David intoned his dirge over Saul and his son Jonathan. 18. He ordered the Judites to be taught The Song of the Bow. It is recorded in the Book of Yashar* (* cf. Josh.10:13; presumably a collection of War Songs):
>
> Your glory, O Israel,
> Lies slain on your heights;
> How have the mighty fallen!
> Tell it not in Gath,
> Do not proclaim it in the streets of Ashkelon,
> Lest the daughters of the Philistine rejoice,
> Lest the daughters of the uncircumcised exult.
> O hills of Gilboah –

Let there be no dew or rain on you,
Or bountiful fields,
For there the shield of warriors lay rejected,
The shield of Saul,
polished with oil no more.
From the blood of slain,
From the fat of warriors –
The bow of Jonathan
Never turned back;
The sword of Saul
Never withdrew empty.
Saul and Jonathan,
Beloved and cherished,
Never parted
In life or death!
They were swifter than eagles,
They were stronger than lions!
Daughters of Israel,
Weep over Saul,
Who clothed you in crimson and finery,
Who decked your robes with jewels of gold.
How have the mighty fallen,
In the thick of battle –
Jonathan, slain on your heights!
I grieve for you,
my brother Jonathan,
You were most dear to me.
Your love was wonderful to me
more than the love of women.
How have the mighty fallen,
The weapons of war perished!" (2 Sam. 1:17-27).

In the ever-changing line-pattern the realistic perception of death is continuously present. Along the theological-anthropological line it is the never loosening grasp of death: 'dust you are and to dust you shall return'; along the literary-metaphorical line it is the intervening power of death in the middle of life; along the personal-emotional line it is the naked fear of death.

FUNERAL CEREMONIES[7]

According to biblical descriptions death was always accompanied by several mourning ceremonies. Garments were torn, shoes taken off, sacks girded up,

7 See in general R.de Vaux, *Hoe het oude Israel leefde* 1, 's Gravenhage 1986², F. "Dood en rouwplechtigheden", 108-117.

heads covered with dust, sometimes the hair was cut and even the bodies were carved. The mourners did not wash themselves and gave up the use of perfumes. Periods of fasting were always observed. However, protests against the ritual, being of excessively pagan origin, are expressed sometimes (Lev.19,27-28; Deut. 14,1).

The pagan uses of necromancy, the consulting of the dead, were forbidden explicitly (Ex.22,18; Lev.20,27; Deut.18,11; Is.8,19). However, on the eve of his decisive battle against the Philistines, Saul, the first king of Israel, had well-founded reasons to consult the dead[8]. "The persecutor of all necromancy has to resort to necromancy himself"[9]. In disguise he applied to "the witch of Endor" and urgently asked for a message from Samuel. The former prophet turned up as an old man cloaked with a mantle and foretold Saul his ultimate fate.

The most important ritual seems to have originated from the loud complaining, uttered by the family and the neighbours. Impressive examples of these mourning-songs, which are of literary quality, have been transmitted in the Hebrew Bible.

Besides the famous lamentation of David after the death of his friend Jonathan (as quoted above), the book Lamentations must be mentioned, functioning as a sorrowful complaint on the occasion of the "death" of Jerusalem, after the conquest by the Babylonians.

To be left unburied in the field or on the mountains was considered a curse. However, not all dead are brought on a bier to the grave, either a simple pit in the earth or a grave hewn in the rock. People of high rank are reported "to be gathered with their fathers"; they had the prerogative to rest in peace in a family-grave, the place of some of which is honoured up to this time. In many graves gifts have been found, either ornaments or jars. The ornaments may have been affixed to decayed garments, whilst the food in the jars may have had a symbolic meaning, functioning as "bread of mourning" and "cup of comfort" (Jer. 16, 7; Ezek. 24, 17-22).

AFTERLIFE

Regarding the issue 'death', the Hebrew Bible does not afford a glance behind the scenes of poems and stories. Within the specific literary context the turns of the metaphor have to be followed. Their imaginative language-use gives an elusive impression of the domains of death, as experienced by men in the midst of life.To draw a balanced scheme or to sketch a conceptual development of textual data may result in mere abstraction which is remote from the text and does not do justice at all to the character of the text.

---

8 The story itself contains sufficient criticism of Saul's behaviour, see Vs. 3 and 9; also 1 Chron.10,13.

9 See K.A.D. Smelik, "The Witch of Endor.1 Samuel 28 in Rabbinic and Christian Exegesis till 800 A.D.", *Vigiliae Christianae* 33 (1977), 160-179.

This general literary state of affairs being the case in the Hebrew Bible, it can hardly be surprising that a distinct word for 'afterlife', or the conceptual notion as such, is lacking. This, however, does not mean that the domain of 'afterlife' or 'future life' was unknown in ancient Israel[10].

The three main parts of the Hebrew Bible, the Torah (the 'Law'), The Ketuvim (the 'Writings') and the Nevi'im (the 'Prophets') make allusions to aspects of the afterlife in their own literary style and particular theological scope.

The lines of two indefinite perceptions of the afterlife may be distinguished. On the one hand, some texts more or less clearly make mention of admission to the realm of God without really dying. Some biblical figures are reported to be suddenly snatched away into heaven, just before their death. On the other hand, several prophetic texts provide examples of types of resurrection of men, some time after their death. With regard to this hopeful outlook, it is not always easy to discern between the individual and the collective-national aspect. Besides, the direct context of the prophetic words is often of considerable influence on their definitive meaning.

The literature of the Ancient Near East as well as the Hebrew Bible of ancient Israel testify to cases of people who came to an untimely end by premature admission[11]. When "Entrückung" is not considered as a provisional but as a permanent admission to the realm of God, some stories or statements in the Hebrew Bible apparently allude to this phenomenon. The first book of the Bible, Genesis, has a unique statement on the ancestral hero Enoch:

> "When Enoch had lived 65 years, he begot Methuselah. After the birth of Methuselah, Enoch walked with God 300 years; and he begot sons and daughters. All the days of Enoch came to 365 years. Enoch walked with God; then he was no more, for God took him" (Gen.5:21-24).

However legendary, prehistoric or historic the statement may be, it somehow or other testifies to the familiarity of an uncommon perception.

The story about the premature 'departure' of the prophet Eliah in the book of Kings is more elaborate. At the approaching farewell to his successor Elisha, both prophets set out from Gilgal and were about to reach the borders of the river Jordan, in the neighbourhood of Jericho:

> "As they kept on walking and talking, a fiery chariot with fiery horses suddenly appeared and separated one from the other; and Eliah went up to heaven in a whirlwind. Elisha saw it, and he cried out, 'Oh, father, father! Israel's chariots and horsemen!' When he could no longer see him, he grasped his garments, and rent them in two" (II Kings 2:1 and 11-12).

---

10 In his study "Hebrew Words for the Resurrection of the Dead" (*Vetus Testamentum* 23, 1973, 218-234), J.F.A. Sawyer only deals with verb-forms ("ordinary, everyday words" and their "special overtones, associations or references which they may have in some contexts", 218); regarding noun-forms he remarks: "The technical term for resurrection of the dead (*techiyyat hametim*) does not occur in Biblical Hebrew, but is attested 4x in the Mishna and 41x in the Talmud", 220.)

11 Cf. Armin Schmitt, *Entrückung-Aufnahme-Himmelfahrt. Untersuchungen zu einem Vorstellungsbereich im Alten Testament* (Forschung zur Bibel 10), Stuttgart 1973.

The stories in the books of Kings, in which Eliah and Elisha as well-known prophets play a dominant role, have as their main religious-historical theme the rivalry between the Lord, the God of Israel, and Baal, the God of the Canaanites. The prophets are pictured as the champions of the Lord enforcing their arguments with a variety of miraculous deeds. The perception and the story of a life without factual dying may have served as a last and best argument in this narrative context[12].

Some texts from the book of Psalms also deserve attention. In 49:16 ("But God will redeem my life from the clutches of Sheol, for He will take me") and 73:24 ("You guided me by your counsel, and led me toward honour") the verses form part of a highly poetic diction and are tightly interwoven with their direct context. Far from being straight statements, these texts in guarded terms first and foremost testify to the believer's confidence and faith in God. This faithful attitude tends to cover lifetime and death's darkness: God's protecting hand will be present, day and night[13].

Finally, some intriguing prophetic words implicitly or explicitly allude to perceptions of the after-life by making mention of some revival from death. Mostly, however, the prophetic references are made in a nationalistic context in view of the eschatological times.
In Isaiah 26 the prophecy:

"Oh, let your dead revive!
Let corpses arise!
Awake and shout for joy,
You who dwell in the dust! – (V. 19)

sounds in direct contrast with a preceding utterance:

"They are dead, they can never live;
Shades, they can never rise" (V. 14).

The people of Vs.19 are "a righteous nation", "a nation that keeps faith" (26:2), whilst Vs.14 contrastingly has in view "those who dwelt high up" (26:5), "Your adversaries" (26:11), that is to say the oppressors of Israel. The enemy will be humbled to the ground (26:5), while Israel, oppressed down to death, will be exalted by his Hand. In high-spirit expectation of the future prophetic language poetically resorts to eloquent images[14].

---

12 Within the Hebrew Bible a close relation of motifs seems to exist with the life story of Moses. Both stories share the miraculous crossing of the river Jordan as well as the mysterious end of life (Deut. 34:5,6). It is no wonder, however, that in post-biblical literature, especially in apocryphal and in pseudepigraphic texts, the mysteries in the lives of the biblical heroes gave ample opportunity to narrative elaborations.

13 Further exegesis and commentary must take account of the wisdom-motifs in the text of both Psalms.

14 "Die Formulierungen, mit denen das Im-Tode-Sein Israels und seine Errettung beschrieben wird, gehen über die traditionellen Formulierungen der Klage- oder Danklieder hinaus. Die unerhörte Situation endzeitlicher Bedrängnis ruft einer Artikulierung in einer außergewönlichen Sprache. Israel steht damit aber an der Schwelle des Glaubens an eine tatsächliche Wiederbelebung nach dem Tode", H.Wildberger, *Jesaja* (*Biblisches Kommentar Altes Testament*, BandX/2, Neukirchen 1978, 996.

In Daniel 12 we find the prophetic announcement:

> "At that time, the great prince, Michael, who stands besides the sons of your people, will appear. It will be a time of trouble, the like of which has never been since the nation came into being. At that time, your people will be rescued, all who are found inscribed in the book. Many of those that sleep in the dust of the earth will awake, some to eternal life, others to reproaches, to everlasting abhorrence" (V.1,2)

This prophecy in great exaltation combines the destiny of the people (V.1) with the destiny of the dead (V.2). The national revival has been brought into comparison with the personal. Metaphorically, both lay down dead, both will awake "at that time", in which the great prince will appear.

With analogous figurative speech Ezekiel 37 unfolds the famous vision of the valley with the "dry bones". Together with his people living in Babylonian captivity, the prophet hears the word of God:

> "He said to me, 'O, mortal, can these bones live again?' I replied, 'O Lord God, only You know'. And He said to me, 'Prophesy over these bones and say to them: O dry bones, hear the word of the Lord! Thus said the Lord God to these bones: I will cause breath to enter you and you shall live again. I will lay sinews upon you, and cover you with flesh, and form skin over you, and I will put breath into you, and you shall live again. And you shall know that I am the Lord" (V. 3-6).
>
> "I prophesied as He commanded me. The breath entered them and they came to life and stood up on their feet, a vast multitude. And He said to me, 'O mortal, these bones are the whole House of Israel" (V.10, 11).

No matter how figurative and suggestive the tale of the vision is, it nevertheless provides with the last verse it own key for understanding in its own context. The House of Israel lives in captivity : "By the rivers of Babylon, there we sat, sat and wept, as we thought of Zion" (Ps. 137:1). The keeping of Jerusalem in memory day and night must have evoked unforeseen images of return and revival, compellingly put into words by Ezekiel, the "mortal", son of Buzi, in the land of the Chaldeans.

In later Jewish and Christian interpretative recontextualization, the vision of the prophet has been understood as the promise and the announcement of the ultimate resurrection at the Last Day[15].

---

15 Cf. H.C.C. Cavallin, "Leben nach dem Tode im Spätjudentum und im frühen Christentum", *Aufstieg und Niedergang der römischen Welt, Geschichte und Kultur Roms im Spiegel der neueren Forschung* II. Prinzipat. Neunzehnter Band (1.Halbband, Religion, Hrsg.W.Haase, Berlin/-New York 1979, 240-345.

BIBLIOGRAPHY

G.W. Anderson, 'Canonical and non-canonical', *The Cambridge History of the Bible*, Vol.I (*From the Beginnings to Jerome*), Cambridge 1970, Part III. The Old Testament, 113-158.

J. Astmann, *Das kulturelle Gedächtnis. Schrift, Erinnerung und politische Identität in frühen Hochkulturen*, München 1992, Zweites Kapitel, Schriftkultur, II.Kanon – zur Klärung eines Begriffes, 106.

C.J. Bleeker, *Het oord van stilte. Dood en eeuwigheid naar oud-Egyptische geloofsbeleving*, Katwijk 1979.

H.C.C. Cavallin, 'Leben nach dem Tode im Spätjudentum und im frühen Christentum', *Aufstieg und Niedergang der römischen Welt, Geschichte und Kultur Roms im Spiegel der neueren Forschung* II. Prinzipat. Neunzehnter Band (1.Halbband, Religion, Hrsg.W.Haase, Berlin/New York 1979, 240-345.

E. Ebeling, *Tod und Leben nach den Vorstellungen der Babylonier*, Berlin 1931

Th. Jacobsen, *The Treasures of Darkness.* History of Mesopotamian Religion, Yale Univ. Press, New Haven and London, 1979[3]

*TANAKH. The Holy Scriptures.*The New JPS Translation According to the Traditional Hebrew Text. The Jewish Publication Society, Phildelphia-New York 1988.

J.F.A. Sawyer, 'Hebrew Words for the Resurrection of the Dead', *Vetus Testamentum* 23 (1973), 218-234.

A. Schmitt, *Entrückung-Aufnahme-Himmelfahrt. Untersuchungen zu einem Vorstellungsbereich im Alten Testament* (Forschung zur Bibel 10), Stuttgart 1973.

K.A.D. Smelik, 'The Witch of Endor.1 Samuel 28 in Rabbinic and Christian Exegesis till 800 A.D.', *Vigiliae Christianae* 33 (1977), 160-179.

J.N. Tromp, *Primitive Conceptions of Death and the Netherworld in the Old Testament* (Biblica Orientalia 21), Rome 1969

R. de Vaux, *Hoe het oude Israel leefde* 1, 's-Gravenhage 1986[2]

H. Wildberger, *Jesaja* (*Biblisches Kommentar Altes Testament*, BandX/2, Neukirchen 1978.

J. Zandee, *Death as an Enemy. Studies in the History of Religion (Supplements to Numen V)*, Leiden 1960

The Soul, Death and the Afterlife in Early and Classical Greece

Jan N. Bremmer

In this century the Western world has seen a meteoric rise of the sciences of psychiatry and psychology: clearly, we all want to care for our soul in this world.[1] However, an early Greek would not have understood this usage of the word 'soul'. In the poems of Homer (ca. 800 BC), the *Iliad* and the *Odyssey*, the word *psyche* has no connection with the psychological side of man whatsoever. Taking this difference from modern ideas as my point of departure, I would like to discuss in this chapter the Greek concept of the soul of the living (1), the Western attitudes towards death (2), the moment of dying (3), the soul of the dead, including reincarnation (4), the underworld (5), and ghosts (6). The direction of my approach will be twofold. On the one hand, I will make use of ideas of anthropology in order to understand the Greek idea of the soul. On the other hand, I will apply the insights of modern historians who have shown that even death has a history to the extent that every time and culture approaches death in a way that can be related to other aspects of that society, such as its religion, family relations and technological level.

THE SOUL OF THE LIVING

We start with the soul of the living.[2] The first surprising thing we notice about the use of the word *psyche* in the earliest Greek literature, the poems of Homer, is its near absence. It is only mentioned as part of the living person at times of crisis, but never mentioned when its owner functions normally. For instance, when the embassy of the Greek army beseeches Achilles to suppress his anger and resume fighting, he complains that he has been continually risking his *psyche* (*Il.* IX.322). And when a spear was pulled from the thigh of Sarpedon, one of the allies of the Trojans, "his *psyche* left him and a mist came upon his eyes" (*Il.* V.696). Other passages show that this *psyche* was located in such curious places as the limbs, the chest or even the flank. In no Homeric passage does it have any psychological

---

1 For an extensive, excellent bibliography see now M. Herfort–Koch, *Tod, Totenfürsorge und Jenseitsvorstellungen in der griechischen Antike* (Munich 1992); see also P. Habermehl, "Jenseits, Jenseitsvorstellungen," *Reall. f. Antike und Christentum* 17 (1996?).

2 I stick closely here to Bremmer, *The Early Greek Concept of the Soul* (Princeton 1983) and Bremmer, "Greek and Hellenistic Concepts of the Soul," in *Death, Afterlife, and the Soul*, ed. L. Sullivan (New York and London 1989), 198–204. For a completely different approach see R. Padel, *In and Out of the Mind. Greek Images of the Tragic Self* (Princeton 1992).

connection. We can only say that when the *psyche* has left the body forever, its previous owner dies.[3]

What, then, constituted the psychological make-up of the early Greeks? Reading Homer, one will find that there is not one seat of the psychological attributes of man, but an enormously varied vocabulary. The most important word for the seat of emotions such as friendship, anger, joy and grief, but also denoting emotion itself, is *thymos*,[4] a word still used today to denote a kind of gland. But there are other words as well, such as one for fury (*menos*), one for the act of the mind (*noos*), and the words for kidney, heart, lungs, liver, and gallbladder – all of which are being used to indicate the seat of emotions or the emotions themselves; moreover, these terms are often used in a semantically indistinguishable and redundant way.[5] It thus seems that there is in Homer not one center of consciousness, not a firm idea of an 'I' who decides what we are doing. Whereas we have one word, 'soul', to denote the dimension of human life that is distinguishable from the body and that to a large extent determines the nature of the human being, the early Greeks had a variety of words to denote this dimension.

How can we explain this situation? First, of course, we can look for parallels. It is the great merit of especially Scandinavian anthropologists to have collected large amounts of data to show that most 'primitive' peoples thought that man has two kinds of souls. On the one hand, there is what these scholars call the free soul, a soul which represents the individual personality. This soul is inactive when the body is active; it only manifests itself during swoons, dreams or at death (the experiences of the 'I' during the swoons or dreams are ascribed to this soul), but it has no connections with the physical or psychological aspects of the body. On the other hand, there are a number of body-souls, which endow the body with life and consciousness, but of which none stands for that part of man that survives him after death.

Unfortunately, students of soul–belief in the Ancient Near East have not yet made use of these insights, even though the material in the *Old Testament* calls out for an analysis along these lines.[6] It is clear that the Homeric concept of the soul of the living is closely related to these ideas. Here too we find on the one hand the *psyche*, a kind of free-soul, and on the other the body-souls, *thymos* and all that. The only thing that is really missing in Homer are stories about the soul leaving the body, but such stories about soul-journeys were told of miracle-men in the archaic age. Let me give one example:

---

3 For a close analysis of the Homeric material see now S.D. Sullivan, "A multi–faceted term. *Psyche* in Homer, the Homeric Hymns and Hesiod," *Studi Ital. Filol. Class.* NS 6 (1988), 151–80.

4 See most recently C. P. Caswell, *A study of* Thumos *in Early Greek Epic* (Leiden 1990); S.D. Sullivan, "Person and *thymos* in the poetry of Hesiod", *Emerita* 61 (1993), 15–40.

5 This is rightly argued by T. Jahn, *Zum Wortfeld 'Seele–Geist' in der Sprache Homers* (Munich 1987), even if he overstates his case, cf. the review by S. R. van der Mije, *Mnemosyne* IV 44 (1991), 440–5.

6 Cf. E. Jacob, in *Theol. Wtb. z Neuen Test.* 9 (Stuttgart 1973), 614–29.

> They say that the soul of Hermotimus of Clazomenae (a Greek city on the West coast of present-day Turkey), wandering apart from the body, was absent for many years, and in different places foretold events such as great floods and droughts and also earthquakes and plagues and the like, while his stiff body was lying inert, and that the soul, after certain periods re-entering the body as into a sheath, aroused it. As he did this often, and although his wife had orders from him that, whenever he was going to be in trance (literally: to depart) nobody should touch his 'corpse', neither one of the citizens nor anybody else, some people went into his house and, having moved his weak wife by entreaty, they gazed at Hermotimus lying on the ground, naked and motionless. They took fire and burned him, thinking that the soul, when it should arrive and have no place anymore to enter, would be completely deprived of being alive – which indeed happened. The inhabitants of Clazomenae honour Hermotimus till the present day and a sanctuary for him has been founded into which no woman enters for the reason above given.[7]

This example, to which we could add others, is a clear illustration of a soul that leaves the body during a trance. Moreover, Hermotimus seems to have practised a certain technique of ecstasy because his murderers saw him lying naked: he will have taken off his clothes, just as Siberian shamans changed their clothes before starting their séances. It is typical for the Greek attitude towards women that his death is occasioned by his deceitful wife: evidently, Eve had many daughters.[8]

Having established an analogy between Homeric and 'primitive' soul-belief, we are of course curious to know why these 'less developed' peoples have these ideas. But when we look in the anthropological studies, we will search in vain for any answers. In fact, studies of soul–belief never even seem to ask this question. I have no full answer myself; some factors, however, may perhaps be indicated. Rather strikingly, the dualistic concept of the soul changes when small 'primitive' peoples become incorporated into larger states or when their culture becomes more differentiated. At this stage, the free-soul starts to incorporate the other souls; e.g. in Greece, where this development starts after Homer, *psyche* gradually acquires the qualities of *thymos* and becomes the center of consciousness.[9] In Athenian tragedy, dramatic situations present persons, especially women, whose *psyche* sighs or melts in despair, suffers pangs, or is 'bitten' by misfortune – emotions never associated with the *psyche* in Homer.[10] This development evidently reflected the growth of the private sphere in Athenian society,

---

7 Apollonius, *Mirabilia* 3.

8 For a full discussion of Hermotimus and similar cases, see Bremmer, *Early Greek Concept of the Soul*, 24–53; F. Graf, *Nordionische Kulte* (Rome 1985), 390–5.

9 The development after Homer is discussed by S.D. Sullivan, "The extended use of *psyche* in the Greek lyric poets (excluding Pindar and Bacchylides", *La Parola del Passato* 44 (1989), 241–62.

10 Cf. F. Solmsen, "*Phren, Kardia, Psyche* in Greek Tragedy", in *Greek Poetry and Philosophy*, ed. D. E. Gerber (Chico 1984), 265–74.

which promoted a more delicate sensibility and a greater capacity for tender feelings. In other words, the more 'primitive', dualistic concept of the soul seems to belong to a less regulated, less differentiated way of life, in which people do not have to make that many choices and in which they need to contain their emotions to a lesser extent than in more developed societies. Apparently, the members of these societies do not need a center of consciousness, as life is lived according to steady rules which do not tolerate exceptions, and some of their emotions need not be continuously suppressed. Needless to say, these earlier societies are usually less individualistic. Their members are primarily members of a group, as they are still in Homer, and only in later times do they become more like separate individuals. These short observations do not of course solve the problem of the development in soul belief, but they may indicate the direction in which a solution is to be found.

DEATH

Until very recently, historians did not occupy themselves much with death. The human attitude towards dying and death was apparently considered to be universal and unchanging and therefore left unstudied. While biological death is indeed universal, the attitude towards death and the beliefs connected with death are demonstrably subject to change. As so often in modern historiography, French historians showed the way, especially the maverick Philippe Ariès, a complete outsider in historical France and head of an information center in a research institute on tropical fruit during most of his working life.[11] His trail-blazing *L'Enfant et la Vie familiale sous l'Ancien Régime* (1960) has inspired historians of the family and childhood ever since its publication, and his splendid *L'Homme devant la mort* (1977) daringly surveys the development of the attitudes towards death from the Middle Ages until modern times. I will not tire the reader with a detailed survey of his views but limit myself to presenting an outline of his findings as a background to my remarks on the attitudes towards death in ancient Greece. This will make it easier to understand the origin of my approach and to see why similar views will not be found in the literature on Greek ideas about death and the afterlife before the 1980s.

According to Ariès, in the first stage of the evolution of attitudes towards death in Western Europe, the period of the earlier Middle Ages, death is seen as an existential fact to be dealt with in traditional ways. Priests hardly play a role in the death scene, and the fate of the individual is quietly subordinated to the future of the collectivity – the family in the first place. Death is familiar and does not make people particularly anxious. The life hereafter is not important for the consciousness of the group and is seen more like a kind of sleep. It is only in the later Middle Ages that an increasing concern with the fate of the soul comes to the fore. Priests take the place of relatives at the death-bed, and salvation of the

---

11 See his autobiography *Un historien du dimanche* (Paris 1980).

soul comes to be seen as highly important, but also more individual and dependent on good works.[12]

In the course of the Renaissance, Reformation and Baroque, there is a growing fear of death, which only slightly abates during the Enlightenment, when we start to find a withdrawal of the clergy and a stress on the presence of the family.[13] In *La nouvelle Héloise*, Rousseau shows Julie making her personal unaided peace with her Maker, surrounded by no one but her intimate family; the emphasis now starts to shift from interest in the dying to those left behind. In this time, we also find the first testimonies of atheism, as we shall soon see.[14] This trend accelerates in the nineteenth century when the belief in hell starts to decline but spiritualism rises to enable people to keep contact with their deceased relatives. Finally, in the twentieth century the emphasis during a case of death completely shifts to the bereaved, who now withdraw from the dead person and leave the care of the corpse to an undertaker, a trend giving rise to the grotesque commercialisation of death parodied by Evelyn Waugh in *The Loved One* or Nancy Mitford's *The American Way of Death*. With Lawrence Stone (note 13) we could call this stage 'Forbidden Death', as dying now takes place in a hospital, often far away from the next-of-kin who have also dwindled in number. There is no comfort left of believing in the afterlife, and our stress on the pursuit of personal happiness often impedes us even to mention death to those we love and who are going to die. Mourning becomes less and less acceptable, as we do not want to be reminded of the presence of death in our midst. There is at present a protest rising against this trend, but it is still too early to find out to what extent this reaction affects popular attitudes. In a way, the study of death itself is part of this changing attitude, and it cannot be by chance that precisely since the 1960s historians have dedicated much effort to shedding light on this subject.

There is, then, in Western Europe a development of an attitude that goes from accepting death, via fearing death, to finally concealing death. At the same time, we see a corresponding change of interest in the afterlife. From relative unimportance, it becomes the overwhelming focus of interest, and at the moment, belief in it seems to be gradually disappearing. Can we look at the practice of dying, the concept of the soul of the dead and the idea of the afterlife in Greece in a similar way?

We start with the early Greek attitude towards death.[15] There can be little

---

12 Note, though, that Ariès' analysis insufficiently takes into account differences between various social groups, such as peasants and monks, and neglects the radical changes in the later Middle Ages, cf. A. Borst, *Barbaren, Ketzer und Artisten* (München/ Zürich 1990), 567–98.

13 This part of Ariès is also in need of some revision, cf. L. Stone, *The Past and the Present Revisited* (New York 1987), 393-410.

14 For this period see especially J. McManners, *Death and the Enlightenment* (Oxford 1981).

15 I follow here closely two studies by C. Sourvinou–Inwood: "To Die and Enter the House of Hades: Homer, Before and After", in *Mirrors of Mortality*, ed. J. Whaley (London 1981), 15–39 and "A Trauma in Flux: Death in the Eighth Century and After", in *The Greek Renaissance of the Eight Century B.C.*, ed. R. Hägg (Stockholm 1988), 33–49.

doubt that early Greece comes very close to the first period sketched by Ariès, the 'Tamed or Domesticated Death'. The disguised goddess Athena tells Odysseus' son Telemachus: "death is common to all men, and not even the gods can keep it off a man they love, when the portion of death which brings long woe destroys him" (*Od.* III.236–8). In contemporary mythology, personified death (Thanatos) is the brother of personified Sleep (Hypnos).[16] This appears to be another way to express that death is something natural, and, once it has come, unthreatening. Death, then, is unavoidable, and even the children of the gods, even mighty Heracles, die and go to Hades.

But this attitude does not stay the same during the course of Greek history. Already in the *Odyssey*, we hear of Menelaus and the Dioscuri escaping the inescapable fate of death; in Hesiod (*Works and Days* 157ff) part of the heroic race goes to the Isles of the Blest; and in classical Greece mystery cults were flourishing by promising individuals a better life in the hereafter. As Sophocles says: "Thrice blessed are those mortals who have seen these rites and thus enter Hades: for them alone there is life, for the others all is misery" (fr. 837). Within the time-span of a few centuries, then, there is a complete change of attitude, even though the old ideas did not die.[17] A character in Euripides' play *Hypsipyle* can still say: "One buries children, one gains new children, one dies oneself. Men do take this heavily, carrying earth to earth. But it is necessary to harvest life like a fruit-bearing ear of corn, and that the one be, the other not" (fr. 757).

THE MOMENT OF DYING

It is this development from accepting death without any questioning to concern for the individual existence in the hereafter which reflects itself at various levels of Greek funerary beliefs and practices, as I will show by turning now to the moment of death.[18] But first we have to mention a preliminary question. In our culture, the moment of death is perceived as a great danger, and the thought of this moment is repressed as long as possible. But if death is accepted as one of the facts of life, as it was by the early Greeks, what, then, did they perceive as the best age to die? The words of the Trojan ally Sarpedon may give us a clue (*Il.* XII. 322-8):

> If it were the case that once we escaped from this battle we would live for ever, ageless and immortal, I would not be fighting in the frontline, nor

---

16 Cf. C. Mainoldi, "Sonno e morte in Grecia antica," in *Rappresentazioni della morte*, ed. R. Raffaelli (Urbino 1987), 9–46; H.A. Shapiro, *Personifications in Greek Art* (Kilchberg and Zürich 1993), 132–65.

17 This is rightly maintained, against Sourvinou–Inwood, by I. Morris, "Attitudes Toward Death in Archaic Greece", *Classical Antiquity* 8 (1989), 296–320, but he overstates his case and insufficiently takes into account the whole range of evidence. Nevertheless, it must be conceded that the debate is only just starting on these issues, and a detailed study should also take into account the material and ritual evidence.

18 On dying see also R. Garland, *The Greek Way of Death* (London 1985), 13–20.

> would I send you into the glory-giving battle. But now innumerable fates of death lie in wait for us anyway, and no man can escape them. So let us go, and either we will give some man the glory he prays for, or some man will give it to us.

These words clearly show that the goal of life for the Homeric warrior is everlasting fame – an ideal which, incidentally, we find among many Indo-European warriors.[19] If we combine this goal with another feeling of the archaic period, namely that old age was seen as a kind of decay to be dreaded more than death, the inference presents itself that for the aristocrats the highest ideal was to live gloriously, even if it implied dying young.[20]

This ideal also appears more indirectly in the *Iliad*, where the poet not uncommonly mentions the young widow of a hero or the old father lamenting his slain son(s).[21] The Trojan hero Euphorbos says to Menelaus, who has killed his brother, "Now truly, noble Menelaos, you shall pay for my brother, whom you have slain and make boast of; you have widowed his wife in her new bridal chamber, and brought unspeakable woe and lamentation to his parents; I shall console them if I slay you" (*Il.* XVII.34ff). And of Diomedes we are told, "He went after Xanthus and Thoon, sons of Phaenops, both late-comers; their father was worn with cruel age, and he had no other sons for his possessions after him. Then Diomedes slew them and robbed them of their lives, both of them; to their father he left lamentation and bitter grief, for he received them not alive returning from battle, and distant kin divided his estate" (*Il.* V.152ff). Undoubtedly, these examples are chosen by the poet to enhance the tragic impact of his story, but at the same time we must notice, I think, that he could not have chosen these particular examples if being killed at a youthful age was not a frequently occurring phenomenon in Homer's time.

The poet of the *Iliad* canonised this ideal of dying gloriously, if young, by making it the decisive feature of the myth of the greatest Greek warrior, Achilles. When he was young, his mother Thetis made him choose between a heroic death at Troy or a long life deprived of fame (*Il.* IX.410–5). He chose an early death, and so should all the youths who had to listen to this story. The ideal lasted long enough to appear on the vases of the archaic period which precisely chose to depict the moment in which the corpse of the warrior was laid out for the funeral. Invariably, this warrior is young.[22] The message remains unmistakable:

---

19 See most recently F. Graf, "Religion und Mythologie im Zusammenhang mit Homer," in *Zweihundert Jahre Homer–Forschung*, ed. J. Lataczs (Leipzig and Stuttgart 1991: 331–62), 358f.; C. Watkins, in *Reconstructing Languages and Cultures*, eds. E. Polomé and M. Winter (Berlin and New York 1992), 411–6; E. Campanile, "Zur Vorgeschichte der idg. Dichterformeln," in *Comparative–Historical Linguistics: Indo–European and Finno–Ugric*, eds. B. Brogyanyi and R. Lipp (Amsterdam 1993), 61–71.

20 Cf. S.C. Humphreys, *The Family, Women and Death* (New York 1983), 144-6; J.–P. Vernant, *Mortals and Immortals* (Princeton 1991), 50–91.

21 Cf. J. Griffin, *Homer on Life and Death* (Oxford 1980), 121ff.

22 Cf. G. Ahlberg, *Prothesis and Ekphora in Greek Geometric Art* (Göteborg 1971).

to die when young is the high-point of your life. What a contrast with our own culture where the death of the young – James Dean, Buddy Holly, President Kennedy, John Lennon – is perceived as particularly tragic!

As we already noted, modern dying is characterised by loneliness. Provided that we don't die suddenly, there is a fair chance that we will end up in a sterile hospital ward, surrounded by machines and with, at the most, one or two relatives present. We may, however, expect a different attitude in a culture which accepts death, and that is exactly what we find in Homer. In battle, the dying man says a few words, particularly requesting a proper burial, something that the warring parties often denied to one another, as Achilles' initial treatment of Hector's body shows. At home, the dying ideally took place peacefully, as Andromache's words of mourning to Hector illustrate: "when dying you did not stretch your hands out to me from your bed and you did not speak a word of wisdom for me to remember for ever, crying night and day" (*Il.* XXIV.742-5). Apparently, and interestingly, a man could expect his wife to be at his side and to support him in his last moments. Similarly, Agamemnon (*Od.* XI.424-6) reproaches his wife and murderess Clytaemnestra that "she did not close my eyes nor did she close my mouth with her hands" – a pathetic detail because Clytaemnestra's hands did rather different things to him. Needless to say, but this in passing, we never hear of Greek husbands doing the same service to their wives.

Modern history and literature have provided us with many examples of impressive death-bed scenes. In particular, in 18th century France, where atheism was perhaps gaining the greatest number of adherents, people could ask for the last rites to be administered by the clergy or reject these Catholic ceremonies altogether.[23] In early Greece, however, death–bed scenes are absent, as men mostly died on the battlefield or peacefully at home, as we already saw, and women's death–beds did not receive any attention in male-dominated Greece. But Greeks of later times were less certain about their death, and in Plato's *Republic* one of the characters says: "When a man gets near to the end of his life, he becomes subject to fear and anxiety about what lies ahead. The stories told about people in Hades – that if you commit crimes on earth you must pay for them down below – although they are ridiculed for a while, now begin to disturb a man's *psyche* with the possibility that they might be true" (1.330de). It is only now that we find a death-bed scene in the proper sense of the word. When in 399 Socrates was condemned to death by the Athenians, he refused to escape from his prison. Plato describes in detail the last moments of his teacher in his dialogue *Phaedo* (117–8). On the day that he had to drink the hemlock, his friends came and his wife Xanthippe was led away, "crying hysterically". Before the executioner entered the prison-room, Socrates and his friends engaged in a discussion about the immortality of the soul. But after Socrates had drunk the hemlock, the poison

---

23 In some cases this was left so obscure that historians still debate whether Voltaire died a Christian or not (an uncertainty which may well have been intentional!), cf. McManners, *Death and the Enlightenment*, 234–69.

caused him to lie down on his back, as his legs started to feel heavy. The executioner pinched his feet and asked if Socrates felt this. Socrates said no. Then he pinched his shins and so he showed him, going up in this way, how he was getting cold and numb. "The coldness was spreading about as far as his abdomen when Socrates uncovered his face, for he had covered it up, and said 'Crito we ought to offer a cock to Asclepius. See to it and don't forget'. 'No, it shall be done,' said Crito. 'Are you sure that there is nothing else?' When he asked this, Socrates made no reply, but after a little while he stirred, and when the man uncovered him, his eyes were fixed. When Crito saw this, he closed the mouth and eyes".

Just before his death Socrates covered up his face. The gesture is not uncommon. Diogenes the Cynic was also found dead wrapped up in his cloak, although he never used to sleep in this way; the great Julius Caesar wrapped himself up when Brutus and the other senators butchered him. In Greece, someone wrapped himself up when he felt terribly ashamed. In other words, the covering up is an act of separation, and the dying man in this way separates himself already from the living. In many countries those to be executed are still covered up, but brave men refuse the blindfold because it is seen as an act of cowardness. Surely, it was not seen like that in ancient Greece.[24]

Someone who is dying is situated between two worlds, that of the living and that of the dead. We know from anthropology that people who fall into the interstices of society, liminal people as they are called, are often ascribed magic powers or suspected of nasty crimes. A well-known example are the gypsies who are supposed to know the future. It is therefore not surprising that we ascribe a special meaning to the last words of the dying: the words they speak in this liminal situation must have a special meaning.[25] This is a quality of the dying which can be found all through Greek history and seems to be independent of the changing thoughts about the afterlife; we, too, are still curious about someone's last words. Homer already uses the belief to let Patroclus forecast to Hector his death and Hector in turn to Achilles. The belief that swans sing their most beautiful song when they feel they are going to die is explained by Plato (*Phaedo* 84E) in this way: they sing so nicely because they already see the beautiful life ahead of them.[26]

---

24 The custom was more widespread than is usually realized, cf. Aristotle, *Rhet.* 2.6.1385a9; Pausanias 4.18.6; Diogenes Laertius 6.77 (Diogenes); Procopius, *Pers.* 2.9; A. Henrichs, in *Entretiens Hardt* 27 (Geneva 1981), 191. Caesar: according to Valerius Maximus 4.5, this happened for the sake of decency.

25 For a fascinating interpretation of Socrates' last words in this direction, see G. Most, "'A Cock for Asclepius'," *Classical Quarterly* 43 (1993, 96–111), 108.

26 For more examples in Greek literature of dying bringing precognition, see R. Janko, *The Iliad: A Commentary* IV (Cambridge 1992), 420; P. Vidal–Naquet, "Le chant du cygne d'Antigone," in *Sophocle: le texte, les personnages*, eds. A. Machin and L. Pernée (Aix–en–Provence 1993), 285–97.

THE SOUL OF THE DEAD

Curiously, the soul of the dead has not received the same systematic attention from scholars as the soul of the living: only a few comparative studies exist which can guide our way in a methodological approach to this subject.[27] Not surprisingly, these studies have looked at the elements of the soul of the living that survive as a soul of the dead. To give one example, the word for the free soul among the Mordvins, a tribe in Siberia, is called *ört*. Since the soul of the dead is also called *ört*, the conclusion can be drawn that the idea of the soul of the dead was derived generically from the soul of the living. In Greece, we would expect that it was the *psyche* that survived man, as in most cases the free-soul is the soul that survives into the afterlife, and that is exactly what we find. The *thymos* never goes to Hades: it is so closely connected with the body that it disintegrates and disappears with it. When heroes like Patroclus and Hector die, Homer says that their *psyche* went to Hades; the dead in the afterlife are indeed often called *psychai*.[28]

The *psyche*, however, was not the only mode of existence after death; the deceased was also compared to a shadow or presented as an *eidolon*, or 'image', a word that suggests that for the ancient Greeks the dead looked like the living. Yet, this is only true to a limited extent, as the physical actions of the souls of the dead were described in two opposite ways. On the one hand, the Greeks believed that the dead souls moved and spoke like the living; the image of the deceased in the memory of the living plays a major part in this activity. In literature, e.g., we find this idea in book XI of the *Odyssey* where Orion and Heracles are depicted as continuing their earthly activities. On the other hand, the souls of the dead are depicted as being unable to move or to speak properly: when the soul of Patroclus leaves Achilles, he disappears squeaking (*Il.* XXIII.100–1).

The circumstance of death was also of some importance in the formation of ideas about the soul of the deceased. Homer describes the warriors at the entrance to Hades still dressed in their bloody armor. Aeschylus has the *eidolon* of Clytemnestra display her death wounds, and Plato elaborately explains this idea, refining it in a way by adding that the soul also retains the scars of its former existence. On vases, the souls of the dead are even regularly shown with their wounds, sometimes still bandaged. It is an idea which we can understand well. When a person dear to us has died in a gruesome accident, we are advised not to look at his dead body, as his last look may haunt us for the rest of our life.[29]

---

27 For this section, see Bremmer, *Early Greek Concept of the Soul*, 70–124. I have added more recent literature where necessary.

28 A. Henrichs, "Namenlosigkeit und Euphemismus," in *Fragmenta dramatica*, eds. H. Hofmann and A. Harder (Göttingen 1991, 161–201), 186–9, points to the surprising fact that in a few important instances the dead are qualified as *apsychoi*, 'soul–less'.

29 Cf. *Od.* XI.41; Aeschylus, *Eumenides* 103; Plato, *Gorgias* 524f.; Bremmer, *Soul*, 83f.

The idea of the soul of the dead in ancient Greece, then, is influenced by the image of the deceased in the memory of the living, by the circumstances of his death, and by the brute fact of the actual dead body. These ideas were never completely systematized and could occur in one and the same description. Just after his death, for example, Patroclus can be described by Homer (*Il.* XXIII. 66–7) as appearing and speaking to Achilles exactly as he was during his life, but as soon as the contact is over he leaves squeaking like a bat.

Yet, these speaking persons are exceptions to the rule, and when the early Greeks spoke of souls of the dead they referred to them as "the wasted ones", "the outworn ones", or "the feeble heads of the dead". Important here, it seems to me, is the plural. The dead are clearly considered to be an enormous, undifferentiated group. In Sophocles they are compared to a swarm of bees: "Up (from the underworld) comes the swarm of the souls, loudly humming" (fr. 879). With so many visitors, it is not hard to understand that the Lord of the Underworld was called "hospitable"; Aeschylus (*Suppliants* 157; fr. 228) even calls him "the most hospitable Zeus of the dead". Especially, when life in the hereafter is not seen as very important, the dead can be lumped together into one big group.[30] It is only when one's own fate is considered important that it becomes less attractive to be a member of the crowd. That does not mean that the early Greeks were not interested in their own survival. On the contrary, the survival they cared for was the fame on earth within their social group, not in the hereafter. This idea of the dead, then, as an anonymous, countless group perfectly fits the early Greek concept of death as an unavoidable, natural process.

Yet we have already seen that in the course of Greek history the *psyche* becomes more and more the focus of the living person. We find another aspect of the development of *psyche* in some lines of the poet Pindar:

> In happy fate all die a death / that frees from care, and yet there still will linger behind / a living image of life, / for this alone has come from the gods.
>
> It sleeps while the members are active; but to those who sleep themselves /it reveals in myriad visions / the fateful approach / of adversities or delights (fr. 131b Maehler).

Here then the soul is described as a typical free-soul (although it is called *eidolon*), but – and this is totally un-Homeric – it is considered 'divine': "this alone comes from the gods". In other words, an enormous revaluation of the soul. Apparently, in addition to a growing unification, the soul was also considered as being more valuable, more important, and the converging of these two developments can hardly be a coincidence: if the soul is gaining in importance, its fate in the hereafter will also become more important.

---

30 Henrichs (n. 28), 194f.

THE UNDERWORLD

It will not come as a surprise that in the period before the fragment that I just quoted there is a growing interest in the fate of the soul. Already in the *Odyssey*, Odysseus goes down to the Underworld to speak to some of the ghosts there. But his trip to Hades is only one of a number of descents into the Underworld of which we hear in the archaic period. The great heroes Heracles and Theseus also went down. And one of the greatest myths of ancient Greece, Orpheus and Eurydice, is not mentioned before 600 BC. We cannot fail to note that the theme of the singer who descends into the Underworld must have been especially attractive to an age where people started to become more interested into their own fate after death, even when the ideas about the soul itself might not yet have greatly changed.[31]

But not only mythical heroes went down to Hades; we also hear of a descent by the man who is perhaps best known by his 'theorem': Pythagoras. Of this curious figure, perhaps the most interesting Greek of the archaic period, who lived in the second half of the sixth century B.C., Hermippus (3rd century BC) tells:

> After arriving in Italy, he built a little underground room and instructed his mother to write down events, as they happened, on a tablet, recording the time, and to keep passing these notes down to him until he came back up. His mother did this, and after some time Pythagoras came up thin as a skeleton. He went into the assembly and announced that he had just returned from Hades. What is more, he read off to them an account of what had happened during his absence. Taken in by his words, they wept and moaned and were sure that he was some kind of divinity; so that they even entrusted their wives to him, thinking that they too would learn something from him. And they were called Pythagorikai (*apud* Diogenes Laertius 8.41, tr. W. Burkert).

This is a most interesting report, as it seems to show that the great philosopher was not above a certain manipulation of his followers – a quality he of course shares with many prophets and miracle-men. What is of interest here for us is that the man who reputedly went down to the underworld had a more than normal interest in the soul. Like various other philosophers and intellectuals of his time, Pythagoras taught the doctrine of reincarnation. Unfortunately, the details of his teachings are most obscure. We do not know whether every living person was supposed to have an immortal soul migrating from one incarnation to the other; whether plants had a soul; whether the soul was immediately incarnated after death, or first rested in Hades as in a kind of purgatory. But independently of all these questions which at the present stage of our knowledge are unanswerable, it

---

31 On Orpheus and Eurydice, see Bremmer, "Orpheus: from guru to gay," in *Orphisme et Orphée*, ed. Ph. Borgeaud (Geneva 1991), 13–30.

is important to note that his doctrine is a testimony to a rising interest in the fate of the soul.[32] Not that everyone shared this interest. A contemporary satirist relates that when Pythagoras saw a dog being beaten, he exclaimed: “Stop! Do not beat him. It is the *psyche* of a dead friend. I recognized him when I heard his whine” (Xenophanes B 7).[33]

To what extent did this revaluation of the soul reflect itself in a revaluation of the underworld? In the *Iliad*, after death, a soul of the dead goes straight to Hades. Evidently, there is no help necessary or any obstacle to prevent the soul from reaching this place. The only thing that the soul has to do is to cross a river, the Styx – a crossing for which no help is required in Homer. The picture of the underworld is bleak and sombre – a gloomy land of no return – and the dead Achilles says rightly: “do not try to make light of death to me; I would sooner be bound to the soil in the hire of another man, a man without lot and without much to live on, than be ruler over all the perished dead” (*Od.* XI.489-91). A similar gloomy idea of the underworld is also one side of the conceptions found in the *Old Testament*, a collection of texts which also testifies to a jest for living and an overall lack of interest in life after death.[34]

But already in late parts of the *Odyssey*, the first indications of a change in this picture become visible. In book XI and XXIV we hear of Hermes as a guide, and in a slightly later poem, the *Minyas* (dating probably from the middle of the sixth century), we hear for the very first time of the ferryman Charon – naturally an old man as the glory of youth would not fit into the gloomy picture.[35] What is the meaning of these guides? There are two aspects here which seem worthwhile to observe. On the one hand, a guide suggests a difficult route. In other words, the appearance of these figures implies that the world of the dead was mentally dissociated from the living: death had apparently become less natural, less easy to tolerate. At the same time, a guide is also someone who knows the way: the need of a reassuring, knowing person is therefore also a sign of a growing anxiety about one’s own fate after death.

This uncertainty reflected itself in an increasing interest in the area of the dead, as illustrated by the accounts of a descent into the underworld, but also in a gradual upgrading of the underworld which in our texts becomes visible in the fifth century. The god of the underworld, who was traditionally called Hades [the Greeks certainly understood this name as ‘invisible one’], now also received the name Pluto or the ‘rich one’, and unlike Hades and Thanatos, Pluto even became the recipient of sacrifices at least from the end of the fourth century. Wealth depended traditionally on agricultural prosperity, and it is therefore not

32 I have underestimated this aspect of reincarnation in Bremmer, *Soul*, 71 note 5.

33 Cf. W. Burkert, *Lore and Science in Ancient Pythagoreanism* (Cambridge Mass. 1972), 156–9 (on Hermippus’ report), 120–3 (reincarnation).

34 For the complicated ideas about afterlife in the *Old Testament*, see most recently K. Spronk, *Beatific Afterlife in Ancient Israel and in the Ancient Near East* (Neukirchen–Vluyn 1986).

35 Cf. C. Sourvinou–Inwood, “Charon,” in *Lexicon Iconographicum Mythologiae Classicae* III.1 (1986), 210–25.

surprising that the dead were also called "Demetreioi", and corn was sown on their graves.[36] This connection of the underworld and material wealth is also expressed by the names used to denote the dead. Whereas in Homer, as we have seen, the dead were preferably called the "powerless heads of the dead", we now find terms such as *olbioi, eudaimones* or *makarioi*. These terms are often translated with 'blessed', but our use of this word is very much influenced by our Christian tradition. The Greeks interpreted these words in a strictly materialistic sense: the dead are the people who are blessed with material goods, who are better off than the living.[37]

In fifth–century Greek comedy, Pherekrates has given a graphic description of the wealth that was awaiting the dead in the beyond (I quote an excerpt):

> All things in the world yonder were mixed with wealth and fashioned with every blessing in every way. Rivers full of porridge and black broth flowed babbling through the channels spoons and all, and lumps of cheese-cake too. Hence the morsel could slip easily and oilily of its own accord down the throats of the dead. Blood-puddings there were, and hot slices of sausage lay scattered by the river banks just like shells. Yes and there were roasted fillets nicely dressed with all sorts of spiced sauces. Close at hand, too, on platters, were whole hams with shin and all, most tender . . . Roast thrushes (...) flew round our mouths entreating us to swallow them as we lay stretched among the myrtles and anemones. And the apples! (...) Girls in silk shawls, just reaching the flower of youth, and shorn of the hair on their bodies, drew through a funnel full cups of red wine with fine bouquet for all who wished to drink. And whenever one had eaten or drunk of these things, straightaway there came forth once more twice as much again (fr. 113 Kassel–Austin, tr. C.B. Gulick, Loeb).

This is a slightly different idea of the hereafter than e.g. in the Islam where, as is well–known, in the afterlife many girls are waiting for the male believers (although in Pherekrates' text a hint of sex is not absent). It is, of course, entirely different from the ideas which were current in early Christianity. When in the *Passio Perpetuae* (202 AD), the future martyr's teacher (himself also to be a martyr) dreams that he has been carried to heaven, he sees "a place whose walls seemed to be constructed of light. And in front of the gate stood four angels, who entered

---

36 Demetrius of Phaleron, fr. 135 Wehrli; Plutarch, *De fac*. 934b.

37 Cf. Henrichs (note 29), 193–8 (Hades and Pluton), 198 (the dead and material wealth). As a matter of trivial curiosity, one of these words is still remembered today, albeit in a rather different meaning. The word *makarios* was taken over in the Christian Church to mean the funeral feast, in later Greek *makaria*. At the same time, the expression was often used together with the word for 'eternal', *aionia*. In the course of time, the two words blended together and formed the name for a favourite Italian dish: macaroni, cf. H. and R. Kahane, *Graeca et Romanica* I (Amsterdam 1979), 400–2.

and put on white robes. We also entered and we heard the sound of voices in unison chanting endlessly: 'Holy, Holy, Holy'" (12.2).
Compared with Pherekrates' vision, this is a rather ascetic idea of the beyond, if completely in harmony with the ascetic life of the early Christians. In Pherekrates, on the contrary, there is an unashamed enjoying of the good things of life. Of course, Old Comedy exploited the idea of the luxurious symposium and banquet in the hereafter, but it occurred also among those groups which were seriously preoccupied with the fate of the soul, such as the Orphics and the Pythagoraeans. Empedocles (B 147), one of the early Greek intellectuals and miracle-men (however curious this combination may seem to us) even stated that having completed its reincarnations the soul would become a table-companion of the gods. This goes back to the mythological idea that important men once shared the table of the gods (Hesiod, fr. 1). This stress on food is of course a reflection of the precarious food situation which would dominate the Western world until this century. There never was a shortage of girls for Greek males – after all, there were brothels galore. But food was always a problem, as it was in the Western Middle Ages, and we cannot fail to see that Pherekrates' picture is closely related to those medieval ones of the Schlaraffenland or the Land of Cockaigne.

GHOSTS

Finally, if the soul of the dead and the afterworld are being revalued, we may also expect a growing interest in contact with ghosts. This is an area which is largely unexplored, and I cannot make more than a few preliminary observations. It seems to me to fit in the development we have sketched that we find a lively interest in ghosts and necromancy, especially in the oldest completely surviving tragedian Aeschylus. Apparitions and conjuring of the dead are important not only in his *Persians* but also in the *Oresteia*, and less than two decades ago a papyrus was published with a scene from his *Psychagogoi* in which a stranger is requested to ask chtonian Zeus to send up from the mouth of the river, surely the Styx, "the swarm of the night-wanderers" (fr. 273a). This seems a new idea. If it is man's destiny to go down to Hades, ghosts of the dead can now come up from the underworld to wander around on earth.[38] We find close contact with the dead also on the vases, where on *lekythoi* it is often impossible to distinguish between the living, as visitors of the tombs, and the dead who have not yet said farewell to the world of the living. In other words, for the surviving family members, there must have existed a close relationship with the dead who were apparently considered to be present in the world of the living, even if limited to the vicinity of the grave. There is not much more to say about the ghosts. In the classical Greek world their presence is still very limited; it is only in later times that necromancy and the subject of ghosts arouse much more interest. But this popularity in later

---

38 Henrichs (note 28), 187–92; add F. Jouan, "L'évocation des morts dans la tragédie grecque," *Revue d'Hist. Rel.* 198 (1981), 403–21.

times can only be understood against the background of an analysis of later attitudes towards death and afterlife. That is, however, as Aristotle would say, the subject of a different inquiry.[39]

39 In addition to being presented to the Amsterdam symposium on death, this paper was read at Princeton (February 1987) and Colgate University (November 1988). I am most grateful to Jan–Maarten Bremer and Theo van den Hout for their careful scrutiny of my text.

LITERARY

Death and Immortality in Some Greek Poems*

Jan Maarten Bremer

There are poets in The Netherlands, a fact which is not widely known outside our country. Among the best and wisest of them is a grand old lady, Elizabeth Eybers. In an interview in a Dutch newspaper[1], she said: "I am surprised at the blackbirds, how every spring they can sing so carelessly and exuberantly. As if they never pay attention to life being so short. But of course they don't. Yeats once said: 'It is man who has invented death.' Man is the only animal who knows he will die. That is miserable – but if we did not know it, life would be much less interesting, and we would not write any poetry, I think."

Readers will understand why I have taken this as an introduction to my theme. "Man has invented death". I take that to mean that man has created all the 'pomp and circumstance' of death; that he has invented countless methods (tricks, spells) to deal with the baffling reality of death. If one looks at what the Greeks did "invent", one finds a broad spectrum from myths which explain why man must die, to rituals which point to life after death; from burial rites and epitaphs which accompany the dead and seal them off in their definitive place in earth, to texts in prose, for example funeral speeches, and in poems which suggest immortality. Poetry plays a fairly central part in all this, and I limit my contribution in this paper to discussing a selection of Greek poems. Taken as a whole, they reflect the above-mentioned polarity between, on the one hand, resignation and endorsement of *death*, even downright denial of immortality: and, on the other, intimations or even certitude of *immortality*. This gives a 'natural' division to my paper.

DEATH

The *Iliad* of Homer is from start to finish concerned with death: the gruesome brutality of killing, the heart-rending reality of dying and the sombre futility of bestowing the last honours upon the dead. Many are the passages in which either the poet himself[2] or his characters reflect upon death; I have selected two for this

---

* I thank David Konstan of Brown University, Providence RI, USA for reading a first draft of this paper and kindly reminding me of some inadequacies in the argument and in the English.

1 NRC-Handelsblad of 25 June 1993.

2 Griffin (1980)103-113 observes that the poet of the *Iliad* in his numerous 'obituaries' brings out the pathos of death. His conclusion (p.143): "The *Iliad* is a poem of death; actual duels in it are short because fate not fighting technique is what interests the poet (...) The lesser heroes, too, are

paper. The first is taken from the words spoken by Sarpedon to Glaukos (12, 322-328). Both are prominent fighters, who have come all the way from Lykia to help the Trojans, their allies. Sarpedon gives two arguments why they should not hesitate to risk their lives on the battlefield. The first is 'noblesse oblige' (310-321), and the second is:

ὦ πέπον, εἰ μὲν γὰρ πόλεμον περὶ τόνδε φυγόντε 322
αἰεὶ δὴ μέλλοιμεν ἀγήρω τ' ἀθανάτω τε
ἔσσεσθ', οὔτε κεν αὐτὸς ἐνὶ πρώτοισι μαχοίμην
οὔτε κέ σε στέλλοιμι μάχην ἐς κυδιάνειραν. 325
νῦν δ' - ἔμπης γὰρ κῆρες ἐφεστᾶσιν θανάτοιο,
μυρίαι, ἅς οὐκ ἔστι φυγεῖν βροτὸν οὐδ' ὑπαλύξαι –
ἴομεν, ἠέ τωι εὖχος ὀρέξομεν, ἠέ τις ἡμῖν.

"Ah my friend, if you and I could escape this fray
and live forever, never a trace of age, immortal,
I would never fight on the front lines again
or command you to the field where men win fame.
But now, as it is, the fates of death await us,
thousands poised to strike, and not a man alive
can flee them or escape – so in we go for attack!
Give our enemy glory or win it for ourselves!"[3]

Sarpedon knows that he and his friend are surrounded by κῆρες μυρίαι, and these κῆρες are no abstractions: in the imagination of the Greeks they are monsters who sweep down on one, ready to swallow him. This gruesome certitude and inevitability of death form an incentive for these heroes to make the most of their life, and to get at least glory out of it – because that is the only thing which will outlive them; but this thought is not expressed here.

The second passage I have chosen, *Iliad* 6, 145-149, has to do with the same Glaukos. When the Greek hero Diomedes sees Glaukos for the first time on the battlefield, he provokes him to reveal whether he is the son of a god, or of a mortal – in the first case Diomedes would rather not fight him. Before giving him the full story about his ancestors (152-211), Glaukos addresses him in the following words:

Τυδείδη μεγάθυμε, τίη γενεὴν ἐρεείνεις; 145
οἵη περ φύλλων γενεή, τοίη δὲ καὶ ἀνδρῶν.
φύλλα τὰ μὲν τ' ἄνεμος χαμάδις χέει, ἄλλα δέ θ' ὕλη
τηλεθόωσα φύει, ἔαρος δ' ἐπιγίγνεται ὥρη·
ὣς ἀνδρῶν γενεὴ ἡ μὲν φύει ἡ δ' ἀπολήγει. 149

---

shown in all the pathos of their death, the grief of their friends and families." -In Vermeule (1979) one finds not only a discussion of this theme (she, too, pays much attention to Homer), but also a wide choice of illustrations (grave-monuments, statues, paintings on pottery) which are very helpful for giving one a sensation of coming close to the Greek experience of death and all that goes with it.

3 *The Iliad of Homer*, translated by Robert Fagles, with an introduction by Robert Knox, Viking-Penguin, New York 1990, pp.335-6

"High-hearted son of Tydeus, why ask about my birth?
Like the generations of leaves, the lives of mortal men.
Now the wind scatters the old leaves across the earth,
now the living timber bursts with the new buds
and spring comes round again. And so with men:
as one generation comes to life, another dies away."[4]

In a more philosophical and less dramatic way, this reflection brings home to us how – according to the Greeks – man is subject to spring and fall, to generation and corruption just like the leaves: mortality is the frame of his existence. This statement does not stand alone: it has been frequently echoed by other Greek poets[5]. Mortality is envisaged here not so much as something inevitable even for the excellent individual, a psychological reality which spurs him on to heroic extremes, but rather as a biological fact of *la condition humaine*.

After the *Iliad*, and strongly influenced by it, comes another body of poetry which is concerned with death even more than the *Iliad* : the poems (there are more than 2000 of them) found on gravestones[6]. Here, too, the material is so rich that I can offer only a few specimens. The first is a text found in Megara on a polyandreion, i.e. a collective tomb. Those who had died in a battle were given a public burial and a tombstone with an inscription, often in verse. The text I have chosen[7] commemorates the men from Megara who had fought and died in the campaigns of the Greeks against the Persians (480-479 BC).

Ἑλλάδι καὶ Μεγαρεῦσιν ἐλεύθερον ἆμαρ ἀέξειν
ἱέμενοι θανάτου μοῖραν ἐδεξάμεθα.

" Wishing passionately to preserve and increase liberty for Hellas and for the people of Megara, we have accepted death as our share." – In all its simplicity this two-liner sheds a clear and painful light upon what has happened. The survivors who have written these lines did not beat about the bush: "we are free, they had to die for it and they had *the courage to accept death*." The verb δέχομαι is of decisive importance here: it highlights the heroic choice of the fighters and gives them the glory about which Sarpedon speaks in the *Iliad.*

My next example[8] is completely different in so far as it is entirely private. It is a poem of an almost naked simplicity, and because of that it has an appeal of its own.

---

4 Fagles, p.200.

5 It is, in fact, the crown-witness for Richard Garner in his *From Homer to Tragedy: the Art of Allusion in Greek Poetry*, Routledge, London New York 1990. On pp. v-vi he shows that allusions to *Iliad* 6,146-9 are found in Mimnermos 2,1-5; Simonides 8,1-2; Bacchylides 5,63-67; Aristophanes *Birds* 685-7; Apollonios Rhodios 4,216-9.

6 From the entire corpus a convenient selection of 500-odd poems has been published by Werner Peek (1960): his *Griechische Grabgedichte* offers, after a thorough introduction (pp.1-42), the Greek texts with a German translation and annotations. Another entry into this area is Lattimore (1942).

7 It is # 6 in Peek's edition; on p.293 one finds interesting details about the inscription as it actually appears on the stone.

8 Peek 40. This epitaph, found on the island of Thasos, is dated shortly after 500 BC.

ἦ καλὸν τὸ μνῆμα πάτηρ ἔστησε θανούσηι
Λεαρέτηι· οὐ γὰρ ἔτι ζῶσαν ἐσοψόμεθα.

" Verily, beautiful is the monument placed for Learete after her death by her father. No, indeed: we shall never again see her alive."

One would imagine from the text that above the inscription a grave-relief preserved the beauty of the girl in stone. This was certainly the case with the following poem[9]:

παιδὸς ἀποφθιμένοιο Κλεοίτου τοῦ Μενεσαίχμου
μνῆμ' ἐσορῶν οἴκτιρ', ὡς καλὸς ὢν ἔθανε.

" This boy Kleoitos passed away, Menesaichmos' son. When you look at his monument, feel sorry for him: how beautiful he was! and even so he died." – It is characteristic for 6th (and 5th) century Athens that the beauty of this boy was not only an object of joy for his admirers and would-be lovers; the boy's father, too, had been proud of it, and it made his death even more poignant.

Completely explicit about the finality of death is this epitaph on a girl[10]:

σάρκας μὲν πῦρ ὄμματ' ἀφείλετο τῆιδε 'Ονησοῦς,
ὀστέα δ' ἀνθεμόεις χῶρος ὅδ' ἄμφις ἔχει.

"Fire took away from our eyes the body of Oneso; here it happened, and now this place, decked with flowers, keeps her bones." – In this burial Oneso's relatives had evidently followed the Homeric fashion (*Iliad* 23, 250-256): the body is cremated, and then the bones are taken out of the ash, piously put into an urn and buried[11].

This small group of epitaphs may suffice to give the reader an impression of the quality of these simple poems in which Homeric metre and Homeric topics were used to express private feelings of loss and sorrow. The Greeks stand alone in this: to my knowledge there are no other languages in which there is such a strong tradition, cultivated through many centuries, of writing and inscribing poems on gravestones. And there is another surprising feature to all this. Poets started writing quasi-epitaphs just for the sake of writing a good short poem. Especially in the Hellenistic period, this became an important literary trend, and I present here two poems by Kallimachos, if not the greatest then certainly the most subtle poet of his century (3rd century BC). The first is a 6-line epigram[12] addressed to a certain Herakleitos from Halikarnassos.

εἶπέ τις, 'Ηράκλειτε, τέον μόρον, ἐς δέ με δάκρυ
ἤγαγεν, ἐμνήσθην δ' ὁσσάκις ἀμφότεροι
ἠέλιον λέσχηι κατεδύσαμεν. ἀλλὰ σὺ μέν που,
ξεῖν' 'Αλικαρνησεῦ, τετράπαλαι σποδίη.

---

9 Peek 45. Found in Athens, dated circa 550 BC.

10 Peek 58 . Found in Athens, dated end 5th century BC.

11 About Homeric burials see Andronikos (1968); about Greek burial customs in general see Kurtz & Boardman (1971).

12 Epigram xxxiv in *Hellenistic Epigrams*, edited by A.S.F. Gow and D.L.Page, Cambridge 1965, vol i, p.65-66.

αἳ δὲ τέαι ζώουσιν ἀηδόνες, ᾗσιν ὁ πάντων 5
ἁρπακτὴς ῎Αιδης οὐκ ἐπὶ χεῖρα βαλεῖ.

"Someone told me, Herakleitos, that you had died, and he made weep, as I was reminded of the many times that the two of us, by our conversations, were the cause of the sun going down in the end. But now, my friend from Halikarnassos, you are ash, four times over. Only your nightingales are alive: on *them* Hades, who snatches away everyone and everything, shall never lay his hand." – One would like to think that this is not a fictitious epitaph but one for a real friend[13]. Two expressions (ἠέλιον κατεδύσαμεν and τετράπαλαι σποδίη) sound colloquial; they seem to refer to casual talk and intimate friendship[14]. For the irretrievable loss there is only one consolation: Callimachus knows that Herakleitos' music, his poems, will live on.

Another poem[15] by Kallimachos is less poignant; in fact its effect and final *frappe* are humorous rather than lugubrious.

– ἦ ρ' ὑπὸ σοὶ Χαρίδας ἀναπαύεται; – εἰ τὸν 'Αρίμμα
τοῦ Κυρηναίου παῖδα λέγεις, ὑπ' ἐμοί.
– ὦ Χαρίδα, τί τὰ νέρθε; – πολὺ σκότος. – αἱ δ' ἄνοδοι τί;
– ψεῦδος. -- ὁ δὲ Πλούτων; - μῦθος. – ἀπωλόμεθα.
– οὗτος ἐμὸς λόγος ὔμμιν ἀληθινός, εἰ δὲ τὸν ἡδύν 5
βούλει, Πελλαίου βοῦς μέγας εἰν 'Αίδηι[16].

"Q. Is it true that Charidas is lying under you? R. If you mean him who was the son of Arimmas from Kyrene[17], yes, he is here right under me. Q. I say, Charidas! how are things down there? R. Damned dark, that's what it is. Q. What about the tales of coming back to the world of the living? R. A pack of lies! Q. And the wealth of Hades? R. That is fiction. Q. Then we are lost, all of us! R. Yes, that is my honest report to you. But if you prefer the delightful version: Here one gets a sirloin steak for a shilling!" (See my note 16).
- We seem to have here an amusing dialogue in which a passer-by puts questions to the grave; of the replies he gets the first one is spoken by the gravestone, the following four by the dead person lying under it. But in fact we have here a

---

13 Actually, there was a poet Herakleitos; one poem (also an funeral epigram, on a young mother) from his hand has survived: *Anth.Pal.*vii, 447; see Gow&Page, vol.i, p.106.

14 A Dutch reader of this poem is reminded of one of the most beautiful poems in his own language: "Egidius, wo bestu bleven? mi lanct na di, gheselle myn" (Egidius, where have you gone? I long for you, my companion).The striking difference is that in the medieval Dutch poem the poet offers himself the consolation that his friend enjoys now the delights of heaven:"noe bestu in den troon verheven, claerre dan der sonnen scyn; alle vrueght esdi gegheven" (now you are elevated on the throne which is more splendid than the brightness of the sun: all joy is given to you). Nothing of that in Kallimachos.

15 Gow & Page xxxi, vol.i, p.65

16 In her *The Secret History* (Knopf, New York 1992, 296), Donna Tartt makes one of her characters use this line 6 of Kallimachos' epigram in Greek: a quotation by which this arrogant student wants to impress and mystify his interlocutor (and Tartt her readers!).

17 Kallimachos himself came from Kyrene, so Charidas and his father may have been real persons known to him.

. debate between the traditional and folkloristic[18] view of death and another ...st cynical view which rejects all illusions. The latter view is in fact the view ...eath which I have been tracing through all these poems, from Homer to the ...ellenistic age; we find it again, phrased with final brevity in the following two-...iner[19] found on a herm in Rome, dated between 150-250 AD.:

οὐκ ἤμην, γενόμην· ἤμην, οὔκ εἰμι· τοσαῦτα.
εἰ δέ τις ἄλλ' ἐρέει, ψεύσεται· οὐκ ἔσομαι.

" I was not – was born – was – am not. That's all.
They tell you something else? Cant! I shan't be."

IMMORTALITY

The second part of this paper will present a selection of Greek poems which – far from accepting, regretting or deploring death in various modalities of resignation – suggest, or even proclaim, immortality. To ease the transition I take my first example again from the poet of whom we have read already two epigrams, Kallimachos[20]:

τῇδε Σάων ὁ Δίκωνος 'Ακάνθιος ἱερὸν ὕπνον
κοιμᾶται· θνήισκειν μὴ λέγε τοὺς ἀγαθούς.

"Here lies Saon, son of Dikon, sleeping a sacred sleep. Do not say that eminent men die."

In itself this simple distichon may be taken to say nothing more than that in the case of ἀγαθοί (people of high courage and virtue) dying, one should refrain from using this heavy and sad word 'die' and speak of 'sleeping' instead. Is this just a euphemism? There is more to it.

In the *Erga* of Hesiod, a poet coming shortly after Homer and at any rate long before Kallimachos, one finds the myth of the five generations: the golden – silver – bronze – heroic – iron ages. Of the men of the golden age, the poet says: θνήισκον δ' ὥσθ' ὕπνωι δεδμημένοι (117): "they died as those who are overcome by sleep", and he goes on (122 ff) to explain that they go on existing on earth (ἐπιχθόνιοι) and act as benevolent powers. The silver generation is slightly worse off than the golden, but they, too, are called blessed μάκαρες (141) after their death. Even the men of the heroic age live on after death: Zeus removes them from the habitat of common men to the isles of the blessed where they live happily: καὶ τοὶ μὲν ναίουσιν ἀκηδέα θυμὸν ἔχοντες, ἐν μακάρων νήσοισι (170-171). It is only in the present, iron age that mortality and all other evils come in at full blast, without any redeeming features.

---

18 The ἄνοδοι (3) refer to the popular narratives about persons who had found a way from Hades back to the upper world: Herakles, Orpheus, Sisyphos, Persephone herself. – Πλούτων (4) is a kind name for Hades, and refers to the rich harvests yielded by the earth. – In line 6 I have tried to find an equivalent for a (probably) proverbial Greek expression which suggests that the realm of the dead is actually a Land of Cockaigne in which food and drink are cheap and plenty. See the commentary of Gow & Page on this line.

19 Peek 453.

20 Gow and Page xli, vol.i p.67.

This mythical frame corresponds to a belief among the Greeks that – even if in their own present life (the iron age) they feel death as inescapable and final – highly eminent people of earlier ages may, in some way, live on and exercise special powers; they called them ἥρωες, heroes[21]. Kallimachos seems to refer to this belief in his epigram on Saon; but there are also many other proofs that brave men who had fallen on the battlefield were given this heroic status by their compatriots. I shall give the most explicit example of this type of epitaph.

In 432 BC, in the military events which would lead to the Peloponnesian War, many Athenian soldiers had fallen before Poteidaia. They were given public burial in Athens, and on their tomb the following poem[22] was written:

ἀθάνατόν με θανοῦσι πολῖται μνῆμ' ἐπέθηκαν,
σημαίνειν ἀρετὴν τῶνδε καὶ ἐσσομένοις
καὶ προγόνων σθένος ἴσον· οἱ ἀντιπάλων πρὸ πόληος
νίκην εὐπόλεμον μνῆμ' ἔλαβον φθίμενοι.
αἰθὴρ μὲν ψυχὰς ὑπεδέξατο, σώματα δὲ χθὼν 5
τῶνδε· Ποτειδαίας δ' ἀμφὶ πύλας ἔλυθεν.

"For the men who have given their lives this immortal monument has been erected by their fellow-citizens, in order to indicate also to future generations the courage of these men and their strength, equal to their forefathers'. They fell before the walls of the enemy city, and they obtained the victory in battle: that was already a monument in itself. The sky has received their souls, and the earth their bodies. It was around the gates of Poteidaia that they were unbound." In line 5 this text[23] tells us that the souls of these heroic men have been taken up high in the air, and this implies that they do not share the corruption (by cremation) of the body; this statement carries, if not an explicit affirmation, at least an intimation of immortality.

It is far from easy to be precise in discerning what the Greeks may have meant by the terms 'soul' and 'immortality'[24]. Greek poets do not theorize, and in their poems one looks in vain for theoretical statements about immortality. There are, however, a number of poetical texts in which in a narrative or lyrical way immortality is suggested; they are almost without exception related to Greek mystery-cults[25], especially those of Demeter and Dionysus.

---

21 W. Burkert (1985), 203-208 discusses the special status of the heroes in the religious imagination of the Greeks.

22 Peek 12. In view of the fact that the thought expressed in lines 5-6 is found in similar words in Euripides' *Suppliants* 533-536 (a play produced by him around 422 BC) and in a fragment preserved from another tragedy (frg.839 Nauck), it cannot be excluded that this epitaph was composed by Euripides who was then at the height of his powers: his *Medea* is of 431.

23 The actual inscription has six more lines; as these are not relevant for my topic I have left them out.

24 For a discussion of the Greek concept of the soul and its (yes or no?) immortality I refer to Bremmer (1983).

25 For the importance and impact of the Greek mystery-cults I refer to Burkert (1985), 276-305 and to Burkert (1987).

An old and venerable text is the *Homeric Hymn to Demeter*[26]. It tells the mythical story of Demeter and her daughter Persephone: Hades, king of the nether world, takes this girl away from the earth, her mother – Demeter, looking everywhere for her daughter, arrives incognito in Eleusis and is well received there by the queen – she is appointed nurse of the queen's son and tries to protect him against death: in vain – Zeus allows Persephone to return to her mother but only for spring and summer – in Eleusis Demeter reveals her identity and gives to its lords the gift of a fertile country and shows them also something sacred:

ἥ δὲ δεῖξε (....) θεμιστοπόλοις βασιλεῦσι
δρησμοσύνην ἱερῶν καὶ ἐπέφραδεν ὄργια πᾶσι,
σεμνά, τά γ' οὔ πως ἔστι παρεξίμεν οὔτε πυθέσθαι
οὔτ' ἀχέειν· μέγα γάρ τι θεῶν σέβας ἰσχάνει αὐδὴν.
ὄλβιος ὃς τάδ' ὄπωπεν ἐπιχθονίων ἀνθρώπων· 480
ὃς δ' ἀτελὴς ἱερῶν, ὅς τ' ἄμμορος, οὔ ποθ' ὁμοίων
αἶσαν ἔχει φθίμενός περ ὑπὸ ζόφωι εὐρώεντι.

" To the kings who wield justice she showed how to do the sacred things, and she revealed the rites to them, the awe-inspiring ones, which one can in no way neglect, nor hear about nor speak of. For a mighty fear of the gods keeps <mortals> from speaking. Blessed among the men who live upon the earth is the person who has seen this. But whosoever is not initiated in the rites and has no part in them, he will never have a share of the same things, after his death, in the realm of darkness and decay."

One thing is certain: not only the Athenians (Eleusis was, to use an anachronism, a suburb of Athens) but all Greeks had a immense respect for the Eleusinian mysteries; for they came in great numbers every autumn to become initiated and "to have a part in them". The myth told them that the gods could not protect them from dying; but the ritual showed them that for the initiates there was some sort of happiness after death. It is impossible to be more precise, either about the rites – they were secret – nor about what the Greeks actually believed – they did not tell us. This much is certain, that the hymn presents a goddess who has painfully felt what it is to see a loved one disappear from earth; this goddess is identical with the one who is responsible for the cycle of vegetation, for agriculture and fertility. It is not surprising that from the rites in honour of this Alma Mater, this feeding mother, the Greeks expected something to overcome their fear of death and to keep hope for 'something after death'.

Of the great poet Pindar a saying[27] is preserved:

ὄλβιος ὅστις ἰδὼν κεῖν' εἶσ' ὑπὸ χθόνα·
οἶδε μὲν βίου τελευτάν,

---

26 N. Richardson published an edition and commentary of this text (Oxford 1974).

27 Frg. 137 (ed. Snell-Maehler). Sophokles frg. 837 (ed. Radt) contains the same 'beatitude': ὡς τρισόλβιοι κεῖνοι βροτῶν, οἵ ταῦτα δερχθέντες τέλη μόλωσ' ἐς ῞Αιδου· τοῖσδε γὰρ μόνοις ἐκεῖ ζῆν ἔστι, τοῖς δ' ἄλλοισι πάντ' ἔχειν κακά ("thrice happy those mortals who have seen these mysteries before going to Hades' house; because for them alone there will be *life* down there – for all others there will be complete misery"). Cp. also Isokrates iv *Panegyr.* 28.

οἶδεν δὲ διόσδοτον ἀρχάν. – For once, his words are simple and straightforward: "Blessed is he who has seen those things and then goes beneath the earth; for he knows the end of life, he knows the beginning given by Zeus." The language is allusive, the promise is open-ended. In another text[28] Pindar does not explicitly refer to the "seeing" of the sacred objects in the Eleusinian cult, but here he is much more explicit about the 'quality of life' which is in store for privileged mortals after death:

τοῖσι λάμπει μὲν μένος ἀελίου
τὰν ἔνθαδε νύκτα κάτω,
φοινικορόδοις δ' ἐνὶ λειμώνεσσι προάστιον αὐτῶν
(...) και χρυσοκάρποισι βέβριθε δενδρέοις,
καὶ τοὶ μὲν ἵπποις γυμνασίοισι (τε λαμπροῖς), τοὶ δὲ πεσσοῖς
τοὶ δὲ φορμίγγεσσι τέρπονται,
παρὰ δέ σφισιν εὐανθὴς ἅπας τέθαλεν ὄλβος.

"For them down there the might of sun is shining even while it is night here. They live outside the city among meadows red with roses, and (the country around) is heavy with trees bearing golden fruits. Some of them take delight in horses, others in shining gymnasia; others again enjoy playing draughts or the lyre. Their happiness is flourishing and complete."

Dionysos, a god hardly less mysterious and certainly more complex than Demeter, is for the Greeks another power who has to do with life and death: in his myth and in his cult the two seem sometimes inextricably intertwined. That his cult included initiation not only in the Hellenistic period but also in the archaic and classical period is certain from a saying of the philosopher Herakleitos (fr.B14 DK) and from a story told by Herodotos (iv, 78-80) about Dionysian rites being celebrated in Olbia, a Greek colony in Southern Russia. On the same location, in Olbia, archeologists have recently discovered bone tablets in the area of a sanctuary[29]. These tablets belong to an archeological context dated to the early fifth century; and one of them is inscribed with ΔΙΟΝ.. ΒΙΟΣ ΘΑΝΑΤΟΣ ΒΙΟΣ ... ΑΛΗΘΕΙΑ ... ΔΙΟ ΟΡΦΙΚΟ. These words, however simple, confirm not only that Dion(ysos) was a reality for these Greeks then and there – that we knew already from Herodotos – but also that (1) he had to do with life and death, (2) this was an important religious truth and (3) he had something 'Orphic' about him[30]. Isolated words are no statements, and West (18-19) therefore offers two tentative explanations of the sequence life-death-life which suggest what the beliefs of these Olbian adepts of Dionysos may have been. One is that a mortal has to pass

---

28 Frg. 129 (Snell-Maehler).

29 The actual excavation took place in 1951; the bone tablets with their graffiti were published in Russian in 1978; Martin West was the first to draw the attention of a wider reading public to these data in *Zeitschr. für Papyrologie und Epigraphik* 45(1982), 17-29. As the excavations in Olbia have yielded quite a few of these tablets, West guesses that "they were membership tokens – bone chips symbolizing participation in common sacrifices" (25).

30 A compact exposition about Orphic texts, beliefs and practices is found in Burkert (1985), 296-301.

through the gate of death in order to come out into another form of life: an experience which Medea seems to hold out for her sons – whom she is going to kill herself – in Eur. *Med.* 1038-1039:

ὑμεῖς δὲ μητέρ' οὔκετ' ὄμμασιν φίλοις
ὄψεσθ', ἐς ἄλλο σχῆμ' ἀποστάντες βίου.

"you will pass into another form of life and not see your mother any more". – Or maybe these Olbians thought that what we call life is in fact a stretch of death: the real life of man precedes his earthly existence and is resumed after it: a view also expressed by Euripides[31]:

τίς δ' οἶδεν εἰ ζῆν τοῦθ' ὃ κέκληται θανεῖν,
τὸ ζῆν δὲ θνήισκειν ἐστι;

"Who knows whether life is found in what people call 'death'? Then 'living' would be 'being dead'!"

The following three poetical texts will be the last I shall deal with; in them names and themes which have come up in the second half of this paper will come together. These texts are inscribed upon thin *gold leaves* which were found by archeologists in graves, on the very bodies of deceased persons. These gold leaves come rather early: on archeological grounds they can be dated to a period 450-300 BC. It is well-known that in antiquity, from the 5th century BC to the Roman age, there was a widespread practice of putting tablets in graves: not shining gold leaves but black leaden tablets, *defixiones*, as they are commonly called. These leaden tablets always carry sinister curses or magical spells: dead people are believed to be in contact with the powers of the nether world, and so the living used graves or coffins as a kind of 'mailbox' to send their messages down. As Zuntz observes in his fundamental study[32] about the gold leaves, "lead is shiny like silver when fresh, but it turns black, and is likely to have been chosen for pernicious purposes because of its dark colour and dead heaviness. The analogy to the Gold Leaves is very much *e contrario* : in the case of the *defixiones* the material is lead, not gold: the wording is a curse not a blessing (...) Gold, the bright and imperishable metal, no doubt was chosen to symbolize the perpetuity of life, just as its opposite, the dark and heavy lead, was used to promote destruction and death."

The first gold leaves which came to light were all found in Southern Italy, and for that reason Zuntz was strongly inclined to deny Orphic influence and to attribute these gold leaves to the spiritual atmosphere of Pythagoreanism: it is well-known that Pythagoras, originally from Samos, came to Southern Italy, settled in Croton and gathered pupils around him who were prepared to adopt his mathematics, his ascetic way of life and his doctrine that the souls of men and animals do not disintegrate with the body but begin a new life (metempsychosis): a doctrine which implies the immortality of the soul.

---

31 But in a context we do not know: fr. 833 (Nauck)

32 Zuntz (1971), 277-393. His observations about the significant contrast between the golden leaves and the leaden tablets is found on p. 279 and 285-6.

Again it is Burkert who pointed out, helped by new evidence[33] – which was unknown to Zuntz when he wrote his book – that one should indeed distinguish between the Bacchic, the Orphic and the Pythagorean traditions, while at the same time acknowledging that there is no strict separation, but rather some overlapping. We can leave this to the specialists and give our attention to the poetical texts themselves. The first one[34], found in a grave near Thourioi, runs as follows:

> ἔρχομαι ἐκ καθαρῶν καθαρά, χθονίων βασίλεια,
> Εὐκλῆς Εὐβουλεύς τε καὶ ἀθάνατοι θεοὶ ἄλλοι,
> καὶ γὰρ ἐγῶν ὑμῶν γένος ὄλβιον εὔχομαι εἶμεν.
> ἀλλά με μοῖρ' ἐδάμασσε καὶ ἀστερόπητα κεραυνῶι.
> κύκλου δ' ἐξέπταν βαρυπενθέος ἀργαλέοιο, 5
> ἱμερτοῦ δ' ἐπέβαν στεφάνου ποσὶ καρπαλίμοισι,
> δεσποίνας δ' ὑπὸ κόλπον ἔδυν χθονίας βασιλείας.
> - ὄλβιε καὶ μακάριστε, θεὸς δ' ἔσηι ἀντὶ βροτοῖο. -
> ἔριφος ἐς γάλ' ἔπετον.

" O Queen of those under the earth, and you Eukles, Eubouleus and all other immortal gods, I come from the pure and am pure, for I claim to be your blessed offspring. But fate has overpowered me, and the thunderbolt-thrower with his lightning. However, I have escaped from the wheel of heavy grief and pains; with my swift feet I have attained the garland for which I had been longing, and I have taken refuge in the lap of the Mistress, the Queen of those under the earth. – Blessed and most blissful man, you will be a god instead of a mortal! – I have fallen a kid into the milk."

Many elements in this text seem concrete: wheel, garland, lap, kid-and-milk. They may refer to rituals which the dead person(s)[35] had undergone during their initiation. It is hazardous to claim that we know what precisely the people may have thought when they inscribed this text on the gold leaf. Zuntz assumes that the deceased was a prominent member of the pious community which took care of the burial, and that he[36] had been killed by lightning; they may have interpreted this exceptional death as a privilege which distinguished him from ordinary mortals. At any rate they were convinced that he was on his way to the gods, that he had escaped from something circular (metempsychosis?) and that Persephone (she must be the 'queen') would caress him in her lap. In line 8 they interrupt "his text" to shout their own congratulations! As for the puzzling four

---

33 Burkert (1985), 300. After the first finds in Southern Italy (Thurii, Petelia; later Hipponion) other golden leaves with comparable or almost identical texts were found in Thessaly (Pelinna and Pharsalos), in Crete (Eleutherna) and even in Rome.

34 Zuntz (1971), 301.

35 Actually this grave contains also the remains of some other persons, with gold leaves inscribed with comparable (but shorter and less eloquent) texts.

36 From the feminine form of the adjective in the first line καθαρά, one might be tempted to think that the deceased was a woman. But the vocatives in line 7 are masculine; probably one has to suppose ψυχά as the noun qualified as καθαρά.

words of line 9, they sound like a colloquial Dutch expression ("ik ben met mijn neus in de boter gevallen"). One could compare the ancient Greek proverb ὄνος εἰς ἄχυρα , "a donkey into the bran"[37]. Evidently this comes from the countryside: when peasants have been winnowing, and have taken the sacks of grain home, it may happen that a stray donkey hits upon the precious heaps of bran, chaff and husks: a paradise for the poor animal! In the same way one might understand "kid in the milk": for lambs and kids there is nothing more important and delicious than sucking and getting as much milk as the mother animal allows them. If a kid fell into a pail of milk, it would be delighted to drink without any restraint[38].

There is a second text found in several variations which M. West[39] has endeavored to combine into an 'archetypus'. This runs as follows:

Μνημοσύνης τόδε θρίον ἐπὴν μέλληισι θανεῖσθαι
ἐν πίνακι χρυσέωι τόδε γραψάτω ἠδὲ φορείτω.

Εὑρήσεις δ' Ἀίδαο δόμων ἐπὶ δεξιὰ κρήνην,
πὰρ δ' αὐτῆι λευκὴν ἑστηκυῖαν κυπάρισσον·
ἔνθα κατερχόμεναι ψυχαὶ νεκύων ψύχονται. 5
ταύτης τῆς κρήνης μηδὲ σχεδὸν ἐμπελάσηισθα.
πρόσθεν δ' εὑρήσεις τῆς Μνημοσύνης ἀπὸ λίμνης
ψυχρὸν ὕδωρ προρέον· φύλακες δ' ἐφ' ὕπερθεν ἔασιν.
οἱ δέ σε εἰρήσονται, ὅ τι χρέος εἰσαφικάνεις.
τοῖς δὲ σὺ εὖ μάλα πᾶσαν ἀληθείην καταλέξαι· 10
εἰπεῖν· Γῆς παῖς εἰμι καὶ Οὐρανοῦ ἀστερόεντος,
αὖταρ ἐμοὶ γένος οὐράνιον· τόδε δ' ἴστε καὶ αὐτοί.
δίψηι δ' εἰμ' αὖος καὶ ἀπόλλυμαι· ἀλλὰ δότ' αἶψα
ψυχρὸν ὕδωρ προρέον τῆς Μνημοσύνης ἀπὸ λίμνης.
καὶ δὴ τοὶ τελέουσι σ' ὑποχθονίωι βασιλείαι, 15
καὶ δὴ τοὶ δώσουσι πιεῖν θείης ἀπὸ κρήνης·
καὶ τότ' ἔπειτ' ἄλλοισι μεθ' ἡρώεσσιν ἀνάξεις.

"This is the leaf of Memory – when one is going to die, one should inscribe this text on a golden tablet and wear it".
"You will find on the right hand side of Hades' dwelling a spring, and, standing by it, a white cypress; there the souls of the dead, when they come down, refresh

37 This expression is found in the comic poet Philemon (fr.188), and was later explained by a grammarian (Apostolius 12.78 in *Paroemiographi Graeci*) ἐπὶ τῶν παρ' ἐλπίδας εἰς ἀγαθὰ ἐμπιπτόντων καὶ τούτοις χρωμένων "used for people who, contrary to their expectation, have met good fortune and are enjoying it to the full."

38 About the 'kid in the milk' see further Zuntz, 326-7, who adduces two other Greek proverbs βοῦς εἰς ἄμητον 'a cow brought into a field just mown' and βατράχωι ὕδωρ 'water for a frog'.

39 "Zum neuen Goldblättchen aus Hipponion', in *Zeitschrift für Papyrologie und Epigraphik* 18 (1975), 229-236. In *Class.Quart.* 34(1984), 89-100, Richard Janko reopens the debate (he gives also a compact bibliography), accepts the principle of West's quest for an original version, and discusses new evidence.

themselves. But do not even approach that spring! More to the front you will find cold water gushing from the lake of Memory. Custodians stand over it. They will ask you to which purpose you have come. You will have to tell them indeed the complete truth: 'I am a son of Earth and of starry Heaven, but my race is from Heaven: you know this yourself as well as I. I am desiccated by thirst and am perishing: come on, give me quickly cold water gushing from the lake of Memory'. – They will actually initiate you into the mysteries of the queen of the nether world, and allow you to drink from the divine spring. And then you will lord it, in the company of the other heroes."

A first thing which surprises is that the opening two lines contain as it were the recipe written above the holy text itself; the person who copied the holy text with a view to her own burial, copied the prescription as well! In the second place: the lines 3ff are a traveller's guide to the netherworld, as we know them from the Egyptian Books of the Dead. The opposition between Lethe[40] and Mnemosyne (=Oblivion and Memory) seems to be of decisive importance. If the soul drinks from Oblivion, everything is lost, probably because then any trace would disappear that one has been initiated, that one is of divine descent and thus has a claim on a happy life after death. In this way the golden leaf is, in our present day speech, a 'ticket to Heaven'. In the third place, this deceased belongs to an elite[41] and should go straight to the spring of Memory – while all other souls go and drink Oblivion. Why? because she is of Titanic origin. In Greek mythology, Ouranos and Gaia are the original divine couple ("in the beginning there was Heaven and Earth" is the point of departure of Hesiod's *Theogony*). They gave life to the Titans, and one of them, Prometheus, moulded man from earth; in another version of the myth mankind came forth from the soot of the lightning-struck Titans. That gives to man a very special position, at least to those who are aware of it, and do not forget it. A last remark: the blessed one will reign with the heroes: here the imagination of the members of this community picks up a term from the mainstream of Greek religion: the special position of 'heroes' after their death.

The last text to be presented in this paper is also the most recent piece of evidence: it was published only six years ago[42]; archeologists found it in a woman's sarcophagus in Pelinna, Thessaly. It is shorter than the two discussed above and is evidently related to the same tradition.

---

40 It is true that Lethe is not actually mentioned in the text; but the polarity between the two springs is evident, and then the spring of lines 3-6 must give the water of Oblivion. The earliest mention of Lethe in the underworld seems to be Plato, *Politeia* 621A.

41 If West's reconstruction of line 15 is correct (*art.cit.* 230), there is also an initiation and thus a *selection* after death, corresponding to the one undergone during one's lifetime.

42 K. Tsantsanoglou and G. Parassoglou,'Two Gold Lamellae from Thessaly', *Hellenika* 37 (1987) 3-16. The sarcophagus can be dated to the 4th cent. B.C. The text is found twice in it: on two ivy-shaped leaves. One of the two omits the lines 4 and 7. On this text see also W. Luppe and R. Merkelbach in *ZPE* 76 (1989),13-16, and Charles Segal in *GrRomByzStud* 31(1990), 411-419.

νῦν ἔθανες καὶ νῦν ἐγένου, τρισόλβιε, ἄματι τῶιδε.
εἰπεῖν Φερσεφόναι σ' ὅτι Βάκχιος αὐτὸς ἔλυσε.
ταῦρος εἰς γάλα ἔθορες.
αἶψα εἰς γάλα ἔθορες.
κριὸς εἰς γάλα ἔπεσες.
οἶνον ἔχεις εὐδαίμονα τιμάν,
κἀπιμένει σ' ὑπὸ γῆν τέλεα ἅσσαπερ ὄλβιοι ἄλλοι.

"Now you have died, and now, thrice blessed, you have been born, on this very day. Tell Persephona that the Bacchic god himself has re-leased you. Bull, you jumped into the milk. Quickly you jumped into the milk. Ram, you fell into the milk. You have got wine, a happy privilege, and below the earth rewards await you, exactly the same as the other blessed ones <enjoy>." – We know by now the actors in this post-mortem drama: Persephone in the first place, then the deceased and the unspecified company of the other happy ones. Dionysos[43], who is not on stage, has authorized the deceased to convey to Persephone an important message: viz. that the god has released or absolved him. Absolved from what? Merkelbach (see my note 42) suggests: "Doch wohl von der Mitschuld an dem 'alten Leid', welches die Titanen der Persephone zugefügt hatten, als sie ihren Sohn Dionysos-Zagreus zerrissen und aßen. Zeus hat die Titanen zur Strafe mit dem Blitz erschlagen, und aus ihrer Asche wurden die Menschen geschaffen, die nun die Schuld der Titanen wie eine Erbsünde mit sich tragen; aber in ihnen lebt auch ein kleiner Teil des Dionysos weiter, eine göttlicher Funke der gestarkt wird, wenn sie die Weihen des Dionysos empfangen und von der Schuld freigesprochen werden." – The 'kid-in-the-milk' (which occupied one line in the 'lightning'-text) is here in three different versions (lines 3-4-5): almost too much of a good thing, one would like to say. – But the most striking element is the first line: "Now you have died and now, thrice blessed, you have been born, on this very day!" Not one of the preceding poems of this second section contained such a hopeful, almost dogmatic statement. It forms a striking contrast with the inscription which I presented as the last one of my first section, and which I translated as: "I was not – was born – was – am not. That's all. They tell you something else? Can't! I shan't be."

This paper can only lead to one conclusion: that the Greeks in various poems spoke with very different voices about 'after death': black nothingness or radiant immortality being the two extremes. Neither they nor we can prove the truth of one of the two voices. Of course this paper was never meant to present such proof. What may have become clear is that about this theme – as about so many

43 West (*ZPE* 18,1975, 234-235) argues that before the 4th century the word βάκχος means nothing more than 'initiate', 'who has undergone a certain purification', and Βάκχιος consequently is an epithet applied to the god in question. But here the 'wine' in line 6, in combination with the *ivy*-shaped golden leaves, leads to identifying this god as Dionysus. – See also S.G. Cole, 'New Evidence for the Mysteries of Dionysus', in *GrRomByzStud* 21(1980), 223-238.

other important themes of human existence: love, beauty, society, war and peace – the Greeks used their imagination and their wit in a surprising way, and that some of them were highly articulate concerning things about which most people can only stammer. Certainly their poems are not the carefree music performed by Elisabeth Eybers' blackbirds; but readers may find that some of these texts, especially the sad ones from the first section, qualify as the sweet nightingales spoken of by Kallimachos.

BIBLIOGRAPHY

M. Andronikos, *Totenkult*, Archeologia Homerica W (Göttingen 1968)
J.N. Bremmer, *The Early Greek Concept of the Soul* (Princeton 1983)
W. Burkert, *Greek Religion: archaic and classical* (Oxford 1985)
W. Burkert, *Ancient Mystery Cults* (Cambridge Mass. 1987)
R. Garland, *The Greek Way of Death* (London 1990)
J. Griffin, *Homer on Life and Death* (Oxford 1980)
D. Kurtz & J. Boardman, *Greek Burial Customs* (Ithaca NY 1971)
R. Lattimore, *Themes in Greek and Roman Epitaphs*, Illinois Studies in Language and Literature 28 (Urbana 1942)
H. Lloyd-Jones, "Pindar and the After-life," in *Pindare*, Entretiens Hardt sur l'antiquité classique 31, ed. O.Reverdin & B. Grange (Geneva 1985)
W. Peek, *Griechische Grabgedichte* (Berlin 1960)
E. Vermeule, *Aspects of Death in Early Greek Art and Literature* (Berkeley 1979)
G. Zuntz, *Persephone. Three Essays on Religion and Thought in Magna Graecia* (Oxford 1971)

Hector's Death and Augustan Politics[1]

Hans Smolenaars

Instead of giving a general description of "Death and the Romans", this contribution concentrates on what is perhaps the most dramatic death in Roman poetry, viz. Turnus' death at the end of the *Aeneid*. The analysis of this death may serve as a case-study of Vergil's intertextual technique to exploit literary tradition (Hector's death) for contemporary politics. Furthermore, it may help to define the differences – as implied by Vergil's text – between the heroic – Homeric – age and sophisticated Augustan society with regard to ideas about warfare in general and killing one's enemy more in particular. Turnus' conduct in the war may seem to deserve death, but his merciless execution by Aeneas as an act of personal revenge does reflect on future society. This philosophical reflection, absent in Homer's epic, is relevant to Augustan politics; life in a peaceful society should be preferred to the cycle of revenge and death.

1

At the end of Vergil's *Aeneid* (written in 29–19 BC), the enemy leader Turnus is killed by Aeneas in a duel arranged to decide the war in Latium between the Trojans and the Italian forces. In the final line of the poem Turnus' soul passes indignant to the Shades below (12.952). The death of a principal character is of course a major constituent of the epic genre, which deals with war and the heroic deeds of its leaders[2]. Since the establishment of generic conventions by Homer (*c.* 750 BC), the narrative of most extant (and complete) Greek and Latin epics builds up to a similar climactic event, after which the tension is resolved by a conciliatory conclusion or a wider perspective. As Farron[3] points out, ancient literary works do not end with a violent or emphatic ending, the *Aeneid* being the only exception to this rule. In Homer's *Iliad*, Vergil's immediate model, Hector is savagely killed by Achilles, but there the violence is followed by lamentations of

---

1 I wish to express my gratitude to Dr Daan den Hengst and Drs Fanny Struyk for commenting upon an earlier draft of this article, to Barbara Fasting for kindly correcting my English at very short notice, to Katja Smolenaars and Ineke Blijleven for editing my manuscript.
The translation of the longer passages from Homer (Rieu, Penguin) and Vergil (Fairclough, Loeb) discussed in this article has been included in the Appendix.

2 Hor. *AP* 73f. *res gestae regumque ducumque et tristia bella / quo scribi possent numero, monstravit Homerus*; *Aen.* 1.1 *arma virumque cano...*

3 S. Farron (1982) 136f.

Hector's family, the funeral of Patroclus, Priam's visit to Achilles, the reconciliation between the Trojan king and the Greek hero and, finally, Hector's funeral. The *Odyssey* ends with Odysseus being reconciled to the relatives of the suitors slaughtered by him and the prospect of peace. Silius' *Punica* and Statius' *Thebaid*, both written at the end of the first century AD in close imitation of Vergil, end with a similar notion of peace.

Vergil's highly exceptional and abrupt ending, which has led to a variety of contradictory interpretations of the *Aeneid* as a whole, actually seems deliberately designed to raise questions about Aeneas' eventual victory. The problem has been extensively studied from the philosophical and moral attitudes of Vergil's own day[4] and from the literary viewpoint of Homeric imitation[5].

In this article I will concentrate on the latter approach, arguing that Vergil deliberately exploited the Homeric 'framing context'[6] for his disturbing conclusion of the *Aeneid*, in order to associate his open, emphatic ending with the 'future' of both *Iliad* (war) and *Odyssey* (peace). This extreme case of intertextuality is closely associated not only with the first beginnings of the Roman empire, but also, as will be argued, with the present dilemma of Augustan politics. Hector's death in the Homeric myth, the recognition of which as Vergil's source is a prerequisite to any interpretation of the *Aeneid*, thus appears to be typologically re-enacted by Vergil in order to raise the matter of Rome's historical future.

2 VERGIL'S AENEID

In his *Aeneid* P. Vergilius Maro (70–19 BC) recounts the journey of his protagonist, son of the goddess Venus and the Trojan prince Anchises, from Troy, captured by the victorious Greeks, to Latium, where his descendants were destined to found Rome. Thus, Vergil's epic is a continuation of Homer's *Iliad*, in which Aeneas has a rather limited role. Now, however, events are presented through the eyes of the defeated Trojans. Vergil's choice of Venus' son Aeneas as the father of the Roman race is in accordance with older traditions, but more important with the descent of the *gens Julia* as proclaimed by Julius Caesar and his adopted son Octavianus/Augustus. This linkage is one among many examples in which Vergil carefully connects the mythical past with Augustus' present reign, which is presented as the τέλος or the ultimate goal of the course of history. The *Pax Augusta* was destined by the Fates and willed by the gods from the beginning.

---

4 Most important P.H. Schrijvers (1978), R.O.A.M. Lyne (1983), Farron (1986), Stahl (1990).

5 G.N. Knauer's excellent study (1964) is the obvious basis for any such approach; for the passages discussed here see also G. Williams (1983), D. West (1990).

6 See C. Martindale, *Redeeming the Text. Latin Poetry and the Hermeneutics of Reception*, Cambridge 1990.

This connection is also established by 'prospective' passages such as Jupiter's prophecy in Book I, covering the 'future' up to 27 BC, and the description of future historical events on Aeneas' shield in book 8, the climax of which is the battle at Actium (31 BC), resulting in the defeat of Antony and Cleopatra. Typological correspondences between e.g. Dido, the tragic queen of Carthage, and Cleopatra, queen of Egypt, and between Aeneas, Romulus and Augustus, also aim at firmly relating the mythological past to the actual events under Octavian's reign and justifying, on the level of the 'objective epic voice'[7], Rome's supremacy over the world; cf. Jupiter's announcement *imperium sine fine dedi* (*Aen.* 1.279).

The story of the *Aeneid* covers a period of 7 years, the greater part of which is told in retrospective by Aeneas to Dido in books 2 (Sack of Troy) and 3 (wandering up to the landing at Carthage). The dramatic action covers the 7th year, starting with the storm by which part of the Trojan fleet was destroyed near Sicily and ending with Turnus' death.

3 TECHNIQUES OF IMITATION

A major characteristic of Vergil's style is the constant and deliberate use of Homer's *Iliad* and *Odyssey* as a framework for Aeneas' wandering (books 1–6, 'Odyssean' part) and the war in Latium (books 7–12, 'Iliadic' part)[8]. This exploitation of the Homeric poems as 'framing context' invites the reader to consider the similarities and, even more important, the differences with the new text, and this often alters his reading of the text. For example, Odysseus' homecoming and his killing of the suitors in the *Odyssey* serves as the structural parallel to Aeneas' obtaining his legal – i.e. ordained by fate – wife Lavinia (~Penelope) and the death of her former fiancé Turnus (~suitors). In order to fully understand Vergil's exploitation of the 'homecoming' theme in the *Odyssey*, it is of importance to realize that Dardanus, founder of Troy, originated from Latium: cf. Apollo's prophecy at Delos (3.94ff.):

> Dardanidae viri, quae vos a stirpe parentum
> prima tulit tellus, eadem vos ubere laeto
> accipiet reduces. antiquam exquirite matrem.[9]

---

7 For a full discussion of Vergil's 'voices' see Lyne (1987).

8 Cf. Macrobius *Sat.* 5.26; Serv. *Aen.* 7.1; Don. *vita Verg.* 1.75 *argumentum varium ac multiplex et quasi amborum Homeri carminum instar.*

9 Apollo's next words, *hic domus Aeneae cunctis dominabitur oris / et nati natorum et qui nascentur ab illis* ('There the house of Aeneas shall lord it over all lands, even his children's children and their race that shall be born of them'), are a direct quote from Poseidon's prophecy to Hera in *Il.* 20.307f. νύν δὲ δὴ Αἰνείαο βίη Τρώεσσιν ἀνάξει, καὶ παίδων παῖδες, τοὶ κεν μετόπισθε γένωνται ('now the great Aeneas shall be King of Troy and shall be followed by his children's

('Ye long-suffering sons of Dardanus, the land which bare you first from your parent stock shall welcome you back to her fruitful bosom. Seek out your ancient mother'.)

We see a striking example of the way the differences between Homer's heroic world and Vergil's new type of hero[10] are highlighted in a comparison between Aeneas' address to the survivors of the storm (*Aen.* 1.198–209) and Odysseus' address in the very same situation (*Od.* 12.208–13). Vergil first establishes the contextual similarity with a 'Leitzitat': *o socii (neque enim ignari sumus ante malorum)* ('O comrades – for ere this we have not been ignorant of evils–') ~ ὦ φίλοι – οὐ γάρ πώ τι κακῶν ἀδαήμονές εἰμεν ('My friends', I said, 'we are men who have met trouble before'). Then both leaders recall earlier calamities overcome, *vos et ... Cyclopea saxa experti* ('Ye drew near to Scylla's fury', 200ff.) ~ οὐ μὲν δὴ τόδε μεῖζον ἔπι κακὸν ἢ ὅτε Κύκλωψ εἴλει ('I cannot see that we are faced here by anything worse than when the Cyclops used his brutal strength to imprison us in his cave', 209f.), which leads to a hopeful conclusion about the future: *revocate animos maestumque timorem / mittite, forsan et haec meminisse iuvabit* ('recall your courage and put away sad fear. Perchance even this distress it will some day be a joy to recall', 202f.) ἀλλά καὶ ἔνθεν ἐμῇ ἀρετῇ βουλῇ τε νόῳ τε ἐκφύγομεν, καί που τῶνδε μνήσεσθαι ὀίω ('Yet my courage and presence of mind found a way out for us even from there; and I am sure that this too will be a memory for us one day', 211). The difference now becomes obvious: Odysseus stresses his excellent capacities as a leader, whereas Aeneas recalls his companions' active part and presents himself as being one of them. This difference is further illustrated by the exhortation concluding both speeches: νῦν δ', ἄγεθ', ὡς ἄν ἐγὼ εἴπω, πειθώμεθα πάντες! ('So now I appeal to you all to do exactly as I say', 213) ~ *durate, et vosmet rebus servate secundis* ('endure, and keep yourselves for days of happiness'), after which Vergil adds the true – concealed – feelings of fearful doubts on the part of his hero: *talia voce refert curisque ingentibus aeger / spem vultu simulat, premit altum corde dolorem* ('So spake his tongue; while sick with weighty cares he feigns hope on his face, and deep in his heart stifles the anguish'). Thus, at the very beginning of the poem the reader is prepared for a completely different type of hero, one ridden with emotions unknown to the Homeric type[11].

---

children'); this is a fine example of Vergil's technique of interweaving his story with elements from Homer's, in this case emphasizing that Aeneas' divine mission and the empire of Rome are firmly rooted in the Homeric past.

10 R.D. Williams (*Virgil*, Greece & Rome, Oxford 1969) 31: "Virgil represents a new type of hero, the 'unheroic' hero who must recognize that the world in which he lives is no longer an heroic world, but one calling for different qualities, less exciting perhaps, and much more difficult to define". For Epicurean influences on Vergil's psychology see M. Erler, 'Der Zorn des Helden. Philodems *De Ira* und Vergils Konzept des Zorns in der *Aeneis*', *Grazer Beiträge* 18 (1992), 103–26.

11 Similar differences, brought about by imitation by means of contrast, have inspired depreciatory judgments such as "Compared with Achilles his Aeneas is but the shadow of a man" (T.E. Page, *The Aeneid of Virgil, I–VI*, 1894: xvii), an interesting reflection of the prejudices of their age.

In his 'Iliadic' books 7–12, Vergil structurally imitates episodes and scenes of the *Iliad* in greater detail.[12] The catalogue in bk. 7 is inspired by *Iliad* 2, the description of Aeneas' shield in bk. 8 transforms Achilles' shield in *Iliad* 18, the fights in bk. 9 during Aeneas' absence follow the structure of those in *Iliad* 8–12, in which Achilles resentfully refrains from battle; the complete structure of books. 10–12 closely imitates the narrative of *Il.* 16–22,336, as will be illustrated below.

Following Knauer's renowned study, we may safely conclude that every episode in Vergil has a model in Homer; but not every Homeric episode has been adopted. The reason for leaving out Homeric scenes is often quite obvious; for instance, large parts of *Od.* 13–24 simply could not be fitted into the narrative because of their specific (Ithacan and domestic) contents. In some cases, however, where long Homeric sequences are closely imitated, I feel we are justified in speculating on the purpose of omissions and transpositions and assuming even that some are *sous–entendue*[13], inviting us to draw conclusions from their absence and to assess the differences with the earlier text. This would apply to a number of famous scenes in the later books of the *Iliad*, which were omitted by Vergil in his detailed structural imitation of these same books and transposed to other parts. For example, the lament for Hector (*Il.* 22.395–515) is omitted from the end of bk. 12, where it structurally belongs, and transposed to the lament for Euryalus in bk. 9[14]. Likewise, the violation of Hector's body has been omitted there and has been included as one of the reliefs on Dido's temple (*Aen.* 2.483ff.), together with the ransom of Hector's body by Priam. Finally, the return of Hector's body in *Iliad* 24 is structurally imitated by the return of Pallas' body to his old father Euander in *Aeneid* 11. In these cases we may assume that a lament for Turnus, the violation of his body, etc., did not suit Vergil's intentions with regard to the conclusion of his poem and that we are expected to recognize and interpret the differences with the earlier text. I will return to this assumption below.

4 BELLA, HORRIDA BELLA

In her obscure and horrifying prophecy in *Aen.* 6.83ff.[15] the Sibyll presents the Italian war (to be narrated by the poet in the 'Iliadic' books 7–12) in terms of the Trojan war:

> o tandem magnis pelagi defuncte periclis
> (sed terrae grauiora manent), in regna Lauini
> Dardanidae uenient (mitte hanc de pectore curam),

12 Knauer (1981) 870ff.

13 Contra West 13 n. 2: "But surely if we assume in a highly imitative literature that omissions are to be *sous–entendue*, chaos will supervene. If an element is omitted in imitation, we should rather assume that it is not wanted"; cf. G. Williams 115.

14 Moreover, as Harrison points out, Achilles picking up the corpse of Hector in *Il.* 24.589 is echoed by Aeneas at 10.830f.

15 See Knauer (1964) 142f. for the structural correspondence with Teiresias' prophecy in *Od.* 11.100–37.

sed non et uenisse uolent. bella, horrida bella,
et Thybrim multo spumantem sanguine cerno.
non Simois tibi nec Xanthus nec Dorica castra
defuerint; alius Latio iam partus Achilles,
natus et ipse dea; nec Teucris addita Iuno
usquam aberit, cum tu supplex in rebus egenis
quas gentis Italum aut quas non oraueris urbes!
causa mali tanti coniunx iterum hospita Teucris
externique iterum thalami.

History will be repeated in the form of a second Simois and Xanthus in Latium, another Achilles (=Turnus), likewise son of a goddess (the nymph Venilia), and again Juno will haunt the Trojans. The cause of this mischief will once more be a wife (Lavinia, a second Helen) and a foreign marriage (Aeneas, a second Paris)[16]. The expectation called up by this prophecy, that Turnus will be cast in the role of Achilles, as he sees himself in *Aen.* 9.148–55, 742, will be done away with in the course of the narrative; his death will be deliberately phrased to recall the death of Hector and Patroclus (see below). Significantly, Vergil did not construct any relation between Aeneas and Hector; rather Troy's greatest hero was chosen to serve as foil to Turnus, Aeneas' antagonist. The only parallel between Aeneas and Hector is provided by his farewell to his son Ascanius (*Aen.* 12.432–40), preceding his duel with Turnus, in imitation of Hector's farewell to Astyanax in *Il.* 6.476–81:

postquam habilis lateri clipeus loricaque tergo est,
Ascanium fusis circum complectitur armis
summaque per galeam delibans oscula fatur:
'disce, puer, uirtutem ex me uerumque laborem,
fortunam ex aliis. nunc te mea dextera bello
defensum dabit et magna inter praemia ducet.
tu facito, mox cum matura adoleuerit aetas,
sis memor et te animo repetentem exempla tuorum
et pater Aeneas et auunculus excitet Hector.'

"Ζεῦ ἄλλοι τε θεοί, δότε δὴ καὶ τόνδε γενέσθαι
παῖδ' ἐμόν, ὡς καὶ ἐγώ περ, ἀριπρεπέα Τρώεσσιν,
ὧδε βίην τ' ἀγαθόν, καί Ἰλίου ἶφι ἀνάσσειν·
καί ποτέ τις εἴποι: 'πατρός γ' ὅδε πολλὸν ἀμείνων'
ἐκ πολέμου ἀνιόντα φέροι δ' ἔναρα βροτόεντα
κτείνας δήϊον ἄνδρα, χαρείη δὲ φρένα μήτηρ.

Hector's prayer to the gods to make his son a valiant warrior becomes a direct exhortation of his son by Aeneas. In *disce virtutem ex me...fortunam ex aliis*, Vergil refers to Soph. *Ai.* 550, ὦ παῖ, γένοιο πατρὸς εὐτυχέστερος, τὰ δε ἄλλ'

16 Line 93 echoes Hector's dying words in *Il.* 22.116 ἥ τ' ἔπλετο νείκεος ἀρχή.

ὁμοῖος· καὶ γένοι' ἄν οὐ κακός[17]. This reference to the misfortunes and depth of Ajax underlines Aeneas' pessimistic view of his own life[18].

Apart from this unique correspondence between Hector and Aeneas, in 7–12 Vergil chose to cast his hero in a different role, as will be illustrated below.

5 *AEN.* 10.445–509 (PALLAS) ~ *IL.* 16.419–505 (SARPEDON), 829–867 (PATROCLUS)

The narrative in *Iliad* 16 successively describes the killing of Sarpedon by Patroclus and of Patroclus by Hector. Both episodes are incorporated by Vergil into his description of Pallas' death at the hands of Turnus in an ill-matched fight (458 *viribus imparibus*)[19].

Situation of the battle
Vergil's imitation of the Sarpedon-episode begins at 10.362ff. where Pallas, son of Euander, makes his first appearance in battle: 365 *ut vidit Pallas* ~ *Il.* 16.419 Σαρπηδὼν δ' ὡς οὖν ἴδ'... Vergil's elaborate description of the nature of the ground, not in Homer, serves to account for the fact that the Arcadian cavalry (Pallas' troops) dismount and attack on foot. The detail is derived from *Il.* 16.426, where Sarpedon jumps off his chariot in keeping with the practice of Homeric warfare. His decision to imitate Homer's passage in great detail forced Vergil to account for the Arcadians fighting on foot by adding the roughness of the ground. He then capitalizes on his addition to the Homeric model, varying "αἰδώς, ὦ Λύκιοι· πόσε φεύγετε; νῦν θοοὶ ἔστε (422) by *quo fugitis, socii ? (...) fidite ne pedibus* (369ff.).

Pallas' prayer before the duel
Pallas prays to Hercules for help in his duel with Turnus, but the god is not allowed to save him from impending death (464f.):

> audiit Alcides iuvenem magnumque sub imo
> corde premit gemitum lacrimasque effundit inanis

The scene is modelled on Zeus' consultation of Hera in *Il.* 16.433ff. presenting to her his dilemma: whether or not to save his son Sarpedon (433–38). Hera brings it home to him that any attempt to save Sarpedon will create a dangerous precedent, since all the gods will want to save their children and there are a great many of them: πολλοὶ γὰρ περὶ ἄστυ μέγα Πριάμοιο μάχονται υἱέες ἀθανάτων (448f.).

---

17 Cf. Pacuvius' adaptation *virtuti sis par, dispar fortunis patris (fr.* 156); Lyne (1987) 8f.

18 Line 440 is self–referential; in *Aen.* 3.343 Andromache asks Aeneas about his son: *ecquid in antiquam virtutem animosque virilis/ et pater Aeneas et avunculus excitat Hector?* When Aeneas points out to his son the example of his own *virtus* and that of Hector, which is in accordance with Roman custom, this is even more dramatic because he is echoing Hector's widow.

19 A study of Harrison's commentary on the passages mentioned is highly recommended.

In Vergil, Hercules is fully aware of his powerlessness, as is clear from the pathetic description of this vigorous god in 464f. Jupiter's consolatory words (467–72) are derived from Hera's warning about establishing precedents in the *Iliad*. Jupiter's thoughts intended to console Hercules, *stat sua quique dies...vitae* (=*Il.* 16.441) and *Troiae sub moenibus/ tot gnati cecidere deum* (=*Il.* 16.448f.) contain, as Harrison (1990) rightly argues, two τόποι of consolation: the universality of death and the *occidet et* motif. But even more interesting is the fact that Jupiter has learnt his lesson, from Hera in Vergil's model-passage. The reference to the Homeric scene is confirmed beyond doubt by the explicit link in 470f. *quin occidit una/ Sarpedon, mea progenies.*

Turnus kills Pallas ~ Patroclus kills Sarpedon
Aen. 10.486–89 is a characteristic adaptation of *Il.* 16.502–505. With *una eademque via sanguis animusque sequuntur* (487) Vergil renders προτὶ δὲ φρένες αὐτῷ ἕποντο· τοῖο δ' ἅμα ψυχήν τε καὶ ἔγχεος ἐξέρυσ' αἰχμήν (504f.). With *ille rapit calidum frustra de vulnere telum* (486) he varies on ἐκ χροὸς ἕλκε δόρυ (504). Not the victor (Patroclus), but the victim (Pallas) is the subject of *rapit*; his action is to no avail: 'by one and the same road (viz. the gaping wound) follow blood and life' (487). Patroclus' brutal act, ὁ δὲ λὰξ ἐν στήθεσι βαίνων (503), will be repeated by Turnus in 495. Vergil adds two further elements: *sonitum super arma dedere* (488) is Homeric (e.g. *Il.* 4.504), but the phrasing recalls Enn. *Ann.* 411 Sk. *sonitum insuper arma dederunt*; with *terram hostilem...petit* (489) he phrases the highly pathetic τόπος of 'dying in a land far away from home'.

From 490 onwards Vergil switches to a second model passage, the killing of Patroclus by Hector in *Il.* 16.829ff. The resulting contamination of the Sarpedon and Patroclus episodes in the description of Pallas' death may have been triggered by Homer's repetition of ὁ δὲ λὰξ ἐν στήθεσι βαίνων ἐκ χροὸς ἕλκε δόρυ (16.503f.) in 862f. δόρυ χάλκεον ἐξ ὠτειλῆς εἴρυσε λὰξ προσβάς[20], imitated by Vergil with *laevo pressit pede* (495). The switchover into Patroclus' death is confirmed by Vergil's structural imitation of Hector's cruel words 830ff. in Turnus' arrogant taunt 491–95 and the spoliation of Pallas' baldric (below). The taunt may well be considered cruel as it involves the parents of the slain man (so Harrison), but Turnus' cruelty is certainly surpassed by Hector, who announces to Patroclus that his corpse will be left unburied to be devoured by vultures (16.836)[21]. The close structural and verbal parallels invite us, first of all, to weigh the differences with the Homeric model passage. Turnus' return of the corpse for burial, although accompanied by a highly sarcastic message to Euander (*qualem*

---

20 Likewise the phrase τέλος θανάτοιο κάλυψε connects the death of Sarpedon (502), Patroclus (855) and Hector (22.361).

21 In *Il.* 22.258 Hector suggests the following terms for the duel: the corpse of the defeated will be returned to his family; Achilles bluntly refuses. In *Il.* 22.335f. he will match Hector's brutality to Patroclus: "So now the dogs and birds of prey are going to maul and mangle you, while we Achaeans hold Patroclus' funeral". For Aeneas' similar conduct when in rage see my note 34.

meruit, Pallanta remitto, 492) is merciful when compared with its immediate model. It is only at a later stage that we are invited to compare Turnus' conduct towards the dead Pallas with that of Aeneas towards Lausus, killed by him with the greatest reluctance in 825ff., which passage is designed to offer a clear contrast between Turnus' gloating sarcasm and Aeneas' *pietas* (826).

The spoliation of Pallas' baldric (*immania pondera baltei)* in 496–500 matches Hector stripping the dead Patroclus and putting on the armour of Achilles in *Il.* 17.125f.[22] Like Turnus, *ovat spolio gaudetque potitus* (500), Hector gloats over his spoils, τὰ μὲν κορυθαίολος Ἕκτωρ αὐτὸς ἔχων ὤμοισιν ἀγάλλεται (18.131f.).

nescia mens hominum ~ οὐ κατὰ κόσμον

For both Hector and Turnus, the act of plundering will be the immediate cause of their death. Not the spoliation itself is reprehensible, as it is a normal act in ancient warfare, but putting on the spoils seems to violate a strict taboo (in Rome)[23]. One is supposed either to dedicate the spoils to a god (as Aeneas with Mezentius' armour in 11.5–11) or to use them for making a generous gift to the dead (as Aeneas with Lausus' armour in 10.827, Achilles with Eëtion's in *Il.* 6.417f.). The authorial intervention in 501–505 condemns in general terms Turnus' act and announces his impending death:

> nescia mens hominum fati sortisque futurae
> et servare modum rebus sublata secundis!
> Turno tempus erit ...

The poet's voice blames Turnus for lack of restraint and moderation, a moral judgment which would seem un–Homeric[24]. But again Vergil echoes Homer; in *Il.* 17.205 Zeus gives an Olympic comment on seeing Hector putting on Patroclus' (= Achilles') arms: τεύχεα δ' οὐ κατὰ κόσμον ἀπὸ κρατός τε καὶ ὤμων εἵλευ. Commentators on Homer give scant attention to Zeus' οὐ κατὰ κόσμον. Griffin[25] having translated lines 201–204 continues with the following statement: "In such passages the poet speaks as it were to himself, and the

---

22 Homer's narrative is slightly inconsistent; in 17.125 Hector spoils the defeated Patroclus, whereas his helmet, spear, shield and cuirass have already been taken away from him by Apollo in 16.793–804.

23 In the *Aeneid* the wearing of a defeated enemy's armour is fatal. Williams (1983:117) points at Euryalus' looted helmet as the immediate cause for his death (*Aen.* 9.365ff.); also Camilla is killed owing to her 'female lust for loot and spoils' (11.782). The swap of weaponry in 2.386ff., used as a stratagem, seems an exception to this rule, but even then it is fatal (411–12).

24 G. Williams 88, 117f. Lyne (1990) 327 states that Vergil condemns "his exultation in triumph", whereas in the *Iliad* Zeus considers Hector's conduct as "inappropriate, not quite in order" since he strips Patroclus of the "divine arms of Achilles". This element of *hybris* is also stressed by Schrijvers 492f. and Harrison (at 501–505), who considers the acts of both Hector and Turnus as "a token of fatal over–confidence".

25 J. Griffin 367.

audience overhears him and shares in his attitude, sympathetic yet free from sentimentality...". However, the sympathy of the audience is perhaps affected by 205, which is not taken into account in Griffin's discussion. I tend to agree with Edwards' interpretation of Zeus' comment as an "implied rebuke for Hector's arrogance in symbolically proclaiming himself the equal of Akhilleus"[26]. But whatever the precise meaning of οὐ κατὰ κόσμον, Vergil obviously intends to link Turnus' ill-advised act with Hector's, once again emphasizing the corresponding relationship set up between his episode and Hector's killing of Patroclus.

In summary, Vergil's description of Pallas' death in bk. 10 is a detailed contamination of *Il.* 16.426–507 (Patroclus killing Sarpedon) and 829–867 (Hector killing Patroclus); Zeus' verdict on Hector's conduct in *Il.* 17.201–206 has been included in the same passage. As for the typological division of roles, it appears that:

- Pallas takes the role of Sarpedon and Patroclus (victims)
- Turnus takes the role of Patroclus and Hector (victors).

Vergil's telescoping of the deaths of Sarpedon and Patroclus allowed him to add the story of Lausus being killed at the hands of Aeneas. This highly original scene, unparalleled in Homer, serves as a significant counterpart to the killing of Pallas, enabling Vergil to sharply contrast Turnus' ruthless conduct with Aeneas' humanity (see Harrison). Moreover, this contamination, while retaining all the important details, is a further demonstration of Vergil's tendency to incorporate Homer as completely as possible. Even more important, the typological identification of Turnus with Hector in this episode prepares us for Aeneas' being cast in the only role that has been so deliberately left for him to fulfil; his grief for Pallas equals that of Achilles for Patroclus and will similarly infuriate him. Thus, the events in bk. 10 were carefully planned as a build-up to the disturbing, yet inevitable conclusion of the poem: Aeneas will take on the role of Achilles.

6 *AEN.* 12.930–52 ~ *IL.* 22.322–66

The narrative structure of Aeneas' fight with Turnus (12.697ff.) closely imitates that of Hector's fight with Achilles, in a slightly different order. The most important correspondences in the narrative which precedes the decisive phase (930ff.) are the following:

– 19–45 King Latinus tries to discourage Turnus from the duel. His vain attempt imitates Priam's efforts to dissuade Hector from confronting Achilles (*Il.* 22.38–76).

– 725–27 Jupiter places the destinies of Turnus and Aeneas on a pair of scales to decide who will die. In the model passage *Il.* 22.208–13 a real decision is made, but in Vergil no result is announced.

---

26 Mark W. Edwards, *The Iliad: a Commentary*, vol. V books 17–20 (Cambridge 1991).

– 742–65 the flight of Turnus, pursued by Aeneas, imitates Homer's chase round the walls of Troy. Vergil adapted his phrasing to the differing situation; in *incertos orbis* (743) he describes the circular movements between Laurentum and the swamps[27]. The chase once more emphasizes the fact that Turnus is fulfilling Hector's role.
– 764f. *neque...praemia, sed Turni de vita* ('for no ... price they seek, but for Turnus' life and blood they strife') imitates *Il.* 22.159f. ἐπεὶ οὐχ ἱερήιον οὐδὲ βοείην ἀρνύσθην (...) ἀλλὰ περὶ ψυχῆς θέον Ἕκτορος ἱπποδάμοιο! ('This was no ordinary race, with a sacrificial beast or a leather shield as prize. They were competing for the life of horse-taming Hector').
– 791–842 Shortly before the final duel Jupiter forbids his wife to obstruct the Trojan victory any longer, in her own interest and that of their marriage. Juno obeys, stipulating that the Trojans and the Rutulians, once they live in peace, will carry the name of Latins and will speak the Latin language: *occidit, occideritque sinas cum nomine Troja* ('fallen is Troy, and fallen let her be, together with her name', 828). In this elegant explanation of the name and language of the Romans, amalgated from Trojans and the local inhabitants of Latium, Vergil incorporates the prospective peace, *pacem...esto*, the marriage of Aeneas with Lavinia (821f.) and the future greatness of Rome, *sit Romana potens Itala virtute propago* ('let be a Roman stock, strong in Italian valour', 827). This structural imitation of the discussion between Zeus and Athena in *Il.* 22.168–87 is confirmed by the similarity of the closing lines:

> 22.186f. ὥς εἰπὼν ὄτρυνε πάρος μεμαυῖαν Ἀθήνην· βῆ δὲ κατ' Οὐλύμποιο καρήνων ἀίξασα.
> ('With which encouragement from Zeus, Athene, who had been itching to play her part, sped down from the peaks of Olympus'.)
>
> 12.841f. adnuit his Iuno et mentem laetata retorsit; interea excedit caelo nubemque relinquit .
> ('Juno assented thereto, and joyfully changed her purpose; meanwhile she passes from heaven, and quits the cloud'.)

In Homer, the scene offers only a limited perspective: peace under Odysseus. Zeus' dilemma – whether or not to save Hector – is answered by Athena echoing Hera's argumentation in *Il.* 16.441–43 (above) with regard to Sarpedon's life. Vergil characteristically avoiding the repetition creates a reconciliation scene between Jupiter and Juno as a substitute. For the first time in the poem Juno is delighted, when Jupiter announces that she will be held in great reverence by the future Romans; *laetata* (841) is a clever transformation of πάρος μεμαυῖαν Ἀθήνην (22.186).
– 843–86 Juturna is forced by divine intervention to abandon her brother Turnus, just as Apollo deserts Hector in *Il.* 22.213.

---

27 West 18f.

– 908–12 As Turnus hurls a huge stone at Aeneas, he is disturbed by *Dira*, the dread goddess sent by Jupiter, and all his efforts fail. His physical and mental state is illustrated by a simile, likening his confusion to being unable to speak or move when in a nightmare:

> ac velut in somnis, oculos ubi languida pressit
> nocte quies, nequiquam avidos extendere cursus
> velle videmur et in mediis conatibus aegri
> succidimus; non lingua valet, non corpore notae
> sufficiunt vires, nec vox aut verba sequuntur
> ('And as in dreams of night, when languorous sleep has weighed down our eyes, we seem to strive vainly to press on our eager course, and in mid effort sink helpless: our tongue lacks power, our wonted strength fails our limbs, nor voice nor words ensue...')

This beautiful simile elaborates on *Il.* 22.199f.:

> ὡς δ' ἐν ὀνείρῳ οὐ δύναται φεύγοντα διώκειν –
> οὔτ' ἄρ' ὃ τὸν δύναται ὑποφεύγειν, οὔθ' ὃ διώκειν –...
> ('It was like a chase in a nightmare, when no one, pursuer or pursued, can move a limb'.)

This dry listing of structural correspondences, while it fails to do justice to Vergil's poetry and originality in handling the Homeric material, may serve to show the poet's deliberate efforts to construct parallels between the decisive duels in the *Iliad* (22) and the *Aeneid* (12).

Turnus' death

In the final scene of the poem Turnus is hit in the thigh (919–29). The wound is not lethal, and he begs Aeneas to spare his life or, in any case, to return his corpse to his family. His entreaty to have pity on his old father Daunus, *fuit et tibi talis/ Anchises genitor* (933f.) recalls both the sarcasm he directs at Euander in 10.492f. and Aeneas' (strongly contrasting) compassion at Lausus' death in 10.810ff.; his phrasing is borrowed from Priam's optimistic assumption that Achilles will understand a father's grief: καὶ δέ νυ τῷ γε πατὴρ τοιόσδε τέτυκται, Πηλεύς ('After all he too has a father of the same age as myself', 22.420f.).

In the Homeric model, Hector is fatally wounded in the throat: ἧ κληῖδες ἀπ' ὤμων αὐχέν' ἔχουσι (22.324f.) and begs Achilles in a very similar manner, ὑπὲρ ... σῶν τε τοκήων, to send his body back to Troy, a favour he himself had refused to Patroclus (16.836).[28] Vergil's departure from the Homeric text is highly significant: for Hector, being fatally wounded, there is no need to beg for his life. Unlike Homer, Vergil purposely creates the opportunity for Aeneas to actually

28 Those in favour of Hector tend to let this go unmentioned, e.g. Griffin 365.

spare Turnus' life.[29] Turnus' complete submission confuses Aeneas (*volvens oculos*) and makes him hesitate for a brief moment. Then, on seeing Pallas' belt, he sinks his sword into Turnus' breast, 'aflame with fury and terrible in anger' (*furiis accensus et ira/ terribilis*, 946f.). This impulsive act of revenge has always astonished readers. In spite of the carefully constructed correspondence to Homer's narrative in Book 22, we expect the outcome of the duel to be different, an expectation which is aroused by the Vergilian differences which are introduced and the prevailing atmosphere of *pietas* in large parts of the poem. Aeneas' sudden brutal act of killing a wounded man who has thrown himself on his mercy strongly affects the interpretation of the poem and is among the main causes of the ambiguity of its political meaning: is it one long encomium on the founding of the Roman empire, justifying its past and present, or an account of the failure and losses, the corruption of power and the madness of the empire? It is impossible here to even sketch the many approaches to this question.[30]

Aeneas' choice seems in the first place at variance with Anchises' exhortation in 6.852f. *parcere subiectis et debellare superbos* ('to spare the humbled, and to tame in war the proud'), and with the Stoic ideals of emotional restraint; swept along by extremes of emotion he loses sight of the high ideal of reconciliation and acts out of a desire for revenge.[31] On the other hand, his conduct is justified by Turnus' breaking of the treaty for the duel with Aeneas (12.324ff.; cf. his *devotio* in 694f.) and Aeneas' moral duty to revenge Pallas' death.[32]

There are many other crucial moments in the epic where Vergil was not free to choose: Creusa must remain behind in Troy, Dido cannot go along to Italy, Palinurus must die at Naples, etc. But here he could have opted for a magnanimous dénouement, a splendid example of the *clementia* propagated by Augustus, *Res Gestae 3: victorque omnibus veniam petentibus civibus peperci. Externas gentes, quibus tuto ignosci potuit, conservare quam excidere malui* ('as victor I

---

29 Schenk 387 is misleading: "Da aber Hektor...nur die Freigabe seines Leichnams...erbittet, tritt das Verlangen des Turnus noch weiter in der Vordergrund". Farron's statement that Turnus "does not ask that his life be spared, only that Aeneas give his corpse to his father (...) Vergil has made Turnus too noble to beg for his life" (1981:97) is likewise biased.

30 An informative survey is provided by W. Suerbaum, *Vergils Aeneis. Beiträge zu ihrer Rezeption im Gegenwart und Geschichte*, Bamberg 1981 (concentrating on the reception of the poem's conclusion in Late Antiquity) and *id.*, 'Hundert Jahre Vergil–Forschung' in *ANRW* 31.1 (1980). A. Wlosok, 'Vergil in der neueren Forschung', *Gymn.* 80 (1973) 129–51, discusses and evaluates the contrasting interpretations of the 'German school' (Vergil justifies the empire and the killing of Turnus) and the Anglo-American/'Harvard school', the latter strongly reflecting the tragedies of recent history (Vietnam): e.g. "We must condemn the sudden rage that causes Aeneas to kill Turnus. The killing...cannot be just: this is beyond doubt expected of us" (K. Quinn, *Vergil's Aeneid* [1968] 273); "It is Aeneas who loses at the end of Book XII" (Putnam 193f.). See further Lyne 1987 and, in defense of Aeneas, Stahl 289f.

31 See Lyne (1990) 335f.

32 Cf. Euander's argumentation in 11.177–80. Donatus' statement is in defence of Aeneas: *ecce servata est in persona Aeneae pietas, qua volebat ignoscere, servata religio Pallanti, quia interfector eius non evasit.*

spared the lives of all citizens who asked for mercy. When foreign peoples could safely be pardoned I preferred to preserve rather than to exterminate them'). Aeneas could have spared Turnus, after which a festive wedding would have concluded the epic. The reconciliation between Jupiter and Juno seemed to prepare for such a happy ending. But Vergil opted for a different conclusion. While he stresses that Turnus submits completely and openly (*vicisti et victum ... Ausonii videre*), renounces his betrothed (*tua est Lavinia coniunx*), and pleads for the cycle of violence to be broken, Vergil bestows on Aeneas the typological role of Achilles, and allows Turnus to die in the role of Hector. To the end, he keeps to his Homeric model, counter to the expectations of readers down through the ages[33]. Tormented by doubts, Aeneas suddenly sees the baldric of Pallas on Turnus' shoulders and is reminded of Pallas' cruel death (*saevi monimenta doloris*). Enflamed by blind fury, he says to Turnus with a snarl that it is because of Pallas that he is about to die, and strikes.

In Homer, Hector – like Turnus – begs for his body to be returned to Troy; Achilles' refusal is cruel: dogs and vultures will devour him (22. 354). Then Hector prophesies Achilles' own death and expires. Achilles responds with sarcasm in τέθναθι! Κῆρα δ' ἐγὼ τότε δέξομαι, ὁππότε κεν δὴ Ζεὺς ἐθέλῃ τελέσαι, ἠδ' ἀθάνατοι θεοὶ ἄλλοι! (365f)). In his otherwise strict imitation of the entire episode, Vergil leaves out both the curt refusal of Achilles and his sarcasm. He has allotted the final element to the atheist Mezentius, who, when his victim Orodes predicts his death, snarls *nunc morere, ast de me divum pater atque hominum rex / viderit* (10.743f.)[34]. These omissions are just as significant as the compelling conclusion imposed by the close parallel between the death of Turnus and the death of Hector. Turnus, presented as *alius Achilles* (6.89), dies like Hector, while in this final scene Aeneas is allotted the typological role of Achilles. Looking back, we find several passages which herald and confirm this identification. In addition to the above–mentioned similarities between Turnus ~ Hector and Aeneas ~ Achilles in XII (Latinus' attempt, Jupiter's scales, the pursuit, the nightmare-effect, Apollo's / Iuturna's departure), we find portents of this identification in:[35]

> 10.517–20 Sulmone creatos / quattor hic iuuenes, totidem quos educat Vfens, / uiuentis rapit, inferias quos immolet umbris / captiuoque rogi perfundat sanguine flammas.

---

33 But cf. Stahl 205: "Repelled by Turnus' unethical, abominable conduct as depicted in Book 10, the attentive reader will join Aeneas in the end in opting for revenge rather than for mercy".

34 In this rage at Pallas' death, Aeneas does not distinguish himself from the cruel Achilles and Mezentius. In 10.530ff. he rejects Magus' offer to pay for his life; in 10.554ff. he refuses Tarquitus a burial (*alitibus linquere feris aut ... piscesque impasti vulnera lambent* ('to birds of prey shalt thou be left; or hungry fish shall suck thy wounds'); in 10.600 he gives a sarcastic answer to Lucagus, who after his brother's death pleads for his own life: *morere et fratrem ne desere frater* ('Die, and let not brother forsake brother!').

35 See L.A. MacKay 11–66; Harrison xxix.

> ('Then, four youths, sons of Sulmo, and as many reared by Ufens, he takes alive, to offer as victims to the dead and to sprinkle the funeral flame with captive blood'.)

These eight young men, to be sacrificed to the dead Pallas, parallel the twelve prisoners of war offered by Achilles to Patroclus in 21.26–33[36].

> In 10.521ff. Magus begs Aeneas to spare his life; Aeneas refuses and kills him:
> 536f. sic fatus galeam laeva tenet atque reflexa / ceruice orantis capulo tenus applicat ensem.
> ('So speaking, he grasps the helmet with his left hand, and bending back the suppliant's neck, drives the sword up to the hilt'.)

This shocking slaughter is a conscious reference to the cruel manner in which Pyrrhus, Achilles' son, kills old Priam:

> 2.552f. implicuitque comam laeva, dextraque coruscum / extulit et lateri capulo tenus abdidit ensem.
> ('and wound his left hand in his hair, while with the right he raised high the flashing sword and buried it to the hilt in his side'.)

The close and deliberate association of Aeneas with Achilles should be disturbing for those who interpret the epic as an ode to Augustus and consider Aeneas to be a Stoic hero.

While in the last Book of the *Iliad* Achilles is given the opportunity to show clemency and compassion to Priam, the *Aeneid* ends abruptly in the blind rage of Aeneas. Indeed the poem offers sufficient indication to enable us to imagine to a certain extent what may follow after this open ending; the domination and marriage of Aeneas, announced in 12.821f., will fulfil Creusa's prediction in 2.783f: *illic res laetae regnumque et regia coniunx / parta tibi* ('There in store for thee are happy days, kingship, and a royal wife'). However, both the continued violence (as after the *Iliad*) and the prospect of peace (as after the *Odyssey*) are conceivable. The latter option is called up by 12.821ff., which refer explicitly to the end of the *Odyssey*: Juno's words *pacem ... (esto) / component, cum ... foedera iungent* are derived directly from Zeus' reassuring words to Athena in *Od.* 24.482ff. ἐπεὶ δὴ μνηστῆρας ἐτίσατο δῖος ' Οδυσσεύς,/ ὅρκια πιστὰ ταμόντες ὅ μὲν βασιλευέτω αἰεί, (...) πλοῦτος δὲ καὶ εἰρήνη ἅλις ἔστω ('Since the admirable Odysseus has had his revenge on the Suitors, let them make a treaty of peace to

36 Aeneas takes these young soldiers prisoner in the heat of battle but does not go back on his decision when the funeral cortege forms (11.81f.). While Homer condemns the conduct of Achilles (*Il.* 23.176), Vergil offers no comment at all. In my view, Aeneas' absence from the ritual sacrifice (11. 96ff.) does not detract from the cruelty of it; but see G. Williams 115f.

establish him as king in perpetuity (...) and let peace and plenty prevail')[37]. Vergil's exact imitation of *Iliad* 16 and 22 forces the reader to take note of the similarities between Hector and Turnus and between the 'Homeric' Aeneas and Achilles, and to follow the parallel to its horrific conclusion. At the same time, he interweaves this unbridled violence with prospects of peace and prosperity after the 'suitor murder' (= the death of Turnus) at the end of the *Odyssey*. In the open ending of the *Aeneid*, the future of the *Iliad* and the *Odyssey* are combined, and the final pieces of the Homeric jigsaw puzzle fall into place.

The exceptionally comprehensive way in which Vergil uses the Homeric epic is a fascinating object of study for those who, with Knauer's help, analyze the complex structure of the *Aeneid*. But the *Aeneid* is more than a literary work of art. The imitation of the *Iliad* shows us the consequences of uncontrolled violence; even Aeneas himself is prey to rampant emotion, so that he loses sight of the common good and the moral ideal of *clementia*. The reference to the 'future' of the *Iliad*, culminating in the destruction of Troy, implies the continuation of this violence after the *Aeneid*. The reference to the end of the *Odyssey*, by contrast, announces the dawn of a peaceful future, following the violent restoration of a just and equitable order. Knauer (1964:327) concludes that Vergil's use of the parallel Turnus/suitors/Hector at the end of his epic opens up the possibility of peace. However, this interpretation does not take into account the strong presence of the *Iliad* in this ending. Hardie[38] sees the abrupt conclusion of the *Aeneid* as a negative treatment of the reconciliation between Priam and Achilles, since in the *Iliad* there is a 'temporary pause in the relentless slaughter'. However, I fail to see why a temporary pause should be considered so much more positive, and the *Odyssey* does not even appear in this interpretation. In my view, by combining the *Iliad* and the *Odyssey* Vergil has deliberately left the 'future' of the *Aeneid* open, although he does indicate the possible outcomes: peace or continued war. Given the nature of the epic, this future is not confined to Aeneas' further life, but extends to actual events taking place in Vergil's day.

7 THE END OF THE *AENEID* AND CURRENT POLITICAL EVENTS

What are the implications of this open ending of the *Aeneid*, in which both future scenarios are interwoven, for the typological relationship between Aeneas and Augustus, between the mythological past and the historical present?

Scholars who take the *Aeneid* as an ode to the empire see the ending as the positively charged climax of the poem: Aeneas' justified action against Turnus disposes of the final barrier to a happy future[39]. In this view, the parallel with

---

37 See Knauer (1964) 322. Zeus' prediction of material prosperity is taken over in Creusa's *res laetae* (2.783).

38 Ph. Hardie 93 ff.

39 Cf. V. Buchheit, *Vergil über die Sendung Roms* (Heidelberg 1963) 105: "von berechtigtem Groll und Kampfeseifer gepackt".

Augustus is unmistakable: his cruel actions as 'Ultor' in the period 42–31 BC were the necessary condition for the *Pax Augusta*. By contrast, the more negative interpretations of Aeneas' 'failure' see the ending as an 'encoded reference to acts of cruelty by Augustus against political rivals'[40]. The 'further voices' theories (Lyne, 1987) attempt to steer a middle course between the 'epic voice' and the 'private voice', which stresses the 'costs of empire' and the individual suffering of victims such as Dido, Turnus and Aeneas.

Aeneas ~ Augustus

I believe that the investigation into the moral, religious and juridical arguments justifying Aeneas' *furor* towards Turnus (Schrijvers, Lyne, Stahl) has demonstrated the ambivalence of this behaviour. In a certain sense, the killing of Turnus is understandable and justified, but it has become charged with highly negative associations as a result of the imitation of Homer analysed above. Historical parallels for such a course present themselves in Octavian's violent actions during the civil wars; one need only think of the execution of 300 senators and *equites* in Perusia in 41 BC *after* their surrender (Suet. *Aug.* 15)[41]. There was considerable disagreement on Augustus' conduct during this period, as we know from Tacitus' account of the various views expressed after his death: *hi* (sc. *dicebant*) *pietate erga parentem et necessitudine rei publicae () ad arma civilia actum, quae neque parari possent neque haberi per bonas artes* ('According to some, "filial duty and the needs of a country () had driven him to the weapons of civil strife – weapons which could not be either forged or wielded with clean hands", *Ann.* I 9.3). This balanced judgement on the goal and the means employed is formulated by the *supporters* of Augustus.

If we consider the ending of the *Aeneid* as a reference to the historical situation in 42–31 BC, then Octavian's actions as 'Ultor' must be assessed in the same way as those of Aeneas: understandable at the time, and according to certain codes even justified, but in the end undesirable and indefensible. Where the future is considered, the choice is open: to continue the violence, as in the 'future' of the *Iliad*, or to break through the fatal circle of revenge and violence (*ulterius ne tende odiis*), as in the 'future' of the *Odyssey*. Seen in this light, the *Aeneid* contains neither an unconditional ode to the empire, nor sharp criticism of Rome's dominion of the world. This is a turning point in history: following a long period

---

40 Gransden 213; Putnam 196 "The progress of empire, as Vergil puts it before the reader, is attributed only to madness, vengeance and death"; Farron (1981) 103 "The pardoning of Turnus would have been the ideal end of the *Aeneid*, if it was meant as pro-Augustan or pro-Roman propaganda".

41 The intriguing study by Powell provides an overview of the acts of violence which Octavianus is reported to have referred to in his autobiography in an effort to defend and justify himself. According to Powell, in Book 10 Vergil is consciously referring to this events; the *Aeneis* is a "defensive project (...) frankly confronting a damaging element in Augustus' image." This intriguing, but fanciful article seems to turn Vergil into an apologist and cunning propagandist for the new regime.

of 'Iliadic' violence, it is now possible – indeed, necessary – to opt for an 'Odyssean' future.

We must then ask ourselves whether, according to this interpretation, Vergil is not sketching a dilemma which has, in fact, already been solved; after Actium the final choice appears to have been made, so that the *Aeneid* (29–19 BC) is retroactively praising Augustus for coming down on the side of peace. Perhaps an ode, after all? Historical material pertaining to the period 31–19 BC is scarce, but there are sufficient grounds to seriously question the stability of the constitution of 27 BC The wars in Spain, Augustus' illness, the trial against M. Primus, proconsul of Macedonia, and the execution of consul Varro Murena on suspicion of conspiracy "revealed the precarious tenure on which the peace of the world reposed"; the year 23 BC "was certainly the most critical, in all the long Principate of Augustus"[42]. Seen against the background of these events, the dilemma described above may still have been extremely topical. After Actium the *Pax Augusta* has by no means been established, as is also clear from Horatius' pessimism in e.g. *Carm.* III 6. The future of Rome is still uncertain.

In my opinion, the *Aeneid* is not a one-sided propagandistic pamphlet, but rather a warning against possible developments. Just as at the end of the *Aeneid*, Augustus and all parties involved are faced with a choice: either peace or the continuation of the cycle of revenge and violence. The typological links between Achilles, Aeneas, and Augustus form the background for this choice; the *Iliad* serves as a warning, while the *Odyssey* offers a hopeful prospect for the future. The imitation of Homer not only serves an aesthetic function; it is also an essential part of the political commitment of Vergil's epic.

---

42 Quotations from R. Syme, *The Roman Revolution* (1939) 333f. See further P.W. Ruikers, *Samenzweringen en Intriges tegen Octavianus Augustus Princeps* (diss. Nijmegen 1966).

APPENDIX

Aen. 1.613–30
Sidonian Dido was amazed, first at the sight of the hero, then at his strange misfortune, and thus her lips made utterance: "What fate pursues thee, goddess–born, amidst such perils? What violence drives thee to savage shores? Art thou that Aeneas, whom gracious Venus bore to Dardanian Anchises by the wave of Phrygian Simois? Yea, I myself remember well Teucer's coming to Sidon; when exiled from his native land, he sought a new kingdom by aid of Belus; my father Belus was then wasting rich Cyprus, and held it under his victorious sway. From that time on, the fall of the Trojan city has been known to me; known, too, thine own name and the Pelasgian kings. Even their foe often lauded the Teucrians with highest praise and would have it that he was sprung from the Teucrians' ancient stock. Come therefore, sirs, and pass within our halls. Me, too, has a like fortune driven through many toils, and willed that at last I should find rest in this land. Not ignorant of ill do I learn to befriend the unhappy.

***Aen.* 6.83–97**
O thou that at last hast fulfilled the great perils of the sea – yet by land more grievous woes await thee – into the realm of Lavinium the sons of Dardanus shall come – relieve thy heart of this care – yet they shall not also joy in their coming. Wars, grim wars I see, and Tiber foaming with streams of blood. A Simois thou shalt not lack, nor a Xanthus, nor a Doric camp. Even now another Achilles is raised up in Latium, he, too, god-

Od. 10.323–35
'Who on earth are you?' she asked. 'What parents begot, what city bred such a man? I am amazed to see you take my poison and suffer no magic change. For never before have I known a man who could resist that drug once he had taken it and swallowed it down. You must have a heart in your breast that is proof against all enchantment. I am sure you are Odysseus, the man whom nothing defeats, the man whom the Giant–slayer with the golden wand always told me to expect here on his way back from Troy in his good black ship. But I beg you now to put up your sword and come with me to my bed, so that in love and sleep we may learn to trust one another.'

dess–born; nor shall Juno anywhere fail to dog the Trojans, whilst thou, a suppliant in thy need, what races, what cities of Italy shalt thou not implore! The cause of all this Trojan woe is again an alien bride, again a foreign marriage! Yield not thou to ills, but go forth to face them more boldly than thy Fortune shall allow thee! Thy path of safety shall first, little as thou deemest it, be opened from a Grecian city.

Aen. 12.433–40
Soon as the shield is fitted to his side, and the corslet to his back, he clasps Ascanius in armed embrace, and, lightly kissing his lips through the helm, he cries: "Learn valour from me, my son, and true toil; fortune from others. Today my hand shall shield thee in war and lead thee where are great rewards: see thou, when soon thy years have grown to ripeness, that thou be mindful thereof, and, as thou recallest the pattern of thy kin, let thy sire Aeneas, and thy uncle Hector stir thy soul!"

Il. 6.472–81
But noble Hector quickly took his helmet off and put the dazzling thing on the ground. Then he kissed his son, dandled him in his arms, and prayed to Zeus and the other gods: 'Zeus, and you other gods, grant that this boy of mine may be, like me, pre–eminent in Troy; as strong and brave as I; a mighty king of Ilium. May people say, when he comes back from battle, "Here is a better man than his father." Let him bring home the bloodstained armour of the enemy he has killed, and make his mother happy.'

Aen. 10.362–75
But in another part, where a torrent had driven far and wide rolling boulders and bushes uptorn from the banks, soon as Pallas saw his Arcadins, unused to charge on foot, turn to flight before pursuing Latium – for the roughness of ground lured them for once to resign their steeds – then, as the one hope in such strait, now with entreaties, now with bitter words, he fires their valour: "Friends, whither flee ye? By your brave deeds I pray you, by your King Evander's name, by the wars ye have won, by my hopes, now springing up to match my father's renown – trust not to flight. 'Tis the

Il. 16.419–26
When Sarpedon saw how his beltless Lycians were falling to Patroclus, son of Menoetius, he told his gallant soldiers what he thought of them. 'For shame, Lycians!' he cried. 'Where are you off to – with such admirable speed? Wait till I meet that fellow over there. I mean to find out who it is that is carrying all before him and has done the Trojans so much harm already, bringing so many of our best men down.'
As he spoke he leapt down from his chariot with all his arms, and on the other side Patroclus, when he saw him, did the same.

sword must hew a way through the foe. Where yonder mass of men presses thickest, there your noble country calls you back, with Pallast at your head. No gods press upon us; mortals, by mortal foes are we driven; we have as many lives, as many hands as they. Lo! ocean hems us in with mighty barrier of sea; even now earth fails our flight; shall we seek the main or Troy?" So speaking, he dashes on into the midst of the serried foe.

Aen. 10.460–73
By my father's welcome, and the board whereto thou camest a stranger, I beseech thee, Alcides, aid my high emprise! May Turnus see me strip the bloody arms from his dying limbs, and may his glazing eyes endure a conqueror!" Alcides heard the youth, and deep in his heart stifled a heavy groan, and shed idle tears. Then with kindly words the Father bespeaks his son: "Each has his day appointed; short and irretrievable is the span of life to all: but to lengthen fame by deeds – that is valour's task. Under Troy's high walls fell those many sons of gods; yea, with them fell mine own child Sarpedon. For Turnus, too, his own fate calls, and he has reached the goal of his allotted years." So he speaks, and turns his eyes away from the Rutulian fields.

Il. 16.431–49
The Son of Cronos of the Crooked Ways saw what was happening and was distressed. He sighed, and said to Here, his Sister and his Wife: 'Fate is unkind to me – Sarpedon, whom I dearly love, is destined to be killed by Patroclus son of Menoetius. I wonder now – I am in two minds. Shall I snatch him up and set him down alive in the rich land of Lycia, far from the war and all its tears? Or shall I let him fall to the son of Menoetius this very day?'
'Dread Son of Cronos, you amaze me!' replied the ox–eyed Queen of Heaven. 'Are you proposing to reprieve a mortal man, whose doom has long been settled, from the pains of death? Do as you please; but do not expect the rest of the immortals to applaud. There is this point, too, that you should bear in mind. If you send Sarpedon home alive, what is to prevent some other god from trying to rescue his own son from the fight? A number of the combatants at Troy are sons of gods, who would resent your action bitterly.

Aen. 10.486–508
In vain he plucks the warm dart from the wound; by one and the same road

Il. 16.502–505
Sarpedon said no more, for Death had descended on his eyes and cut short his

follow blood and life. Prone he falls upon the wound, his armour clashes over him, and, dying, he smites the hostile earth with blood–stained mouth. Then standing over him, Turnus cries: "Arcadians, give heed, and bear these my words back to Evander: even as he has merited, I send him back Pallas! Whatever honour a tomb gives, whatever solace a burial, I freely grant; yet his welcome of Aeneas shall cost him dear." So saying, with his left foot he trod upon the dead, tearing away the belt's huge weight and the story of the crime theron engraved – the youthful band foully slain on one nuptial night, and the chambers drenched with blood – which Clonus, son of Eurytus, had richly chased in gold. Now Turnus exults in the spoil and glories in the winning. O mind of man, knowing not fate or coming doom or how to keep bounds when uplifted with favouring fortune! To Turnus shall come the hour when for a great price he will long to have bought an unscathed Pallas, and when he will abhor those spoils and that day.

breath. Patroclus put his foot on his chest and withdrew the spear from his flesh. The midriff came with it: he had drawn out the spear–point and the man's soul together.

Il. 16.827–36
Menoetius' valiant son fell to a short spear-cast from Hector son of Priam, who now addressed him as a conqueror. 'Patroclus,' he said, 'you thought you would sack my town, make Trojan women slaves, and ship them off to your own country. You were a fool. (...) So now the vultures here are going to eat you up.'

Il. 16.855–63
Death cut Patroclus short and his disembodied soul took wing for the House of Hades, bewailing its lot and the youth and manhood that it left. But illustrious Hector spoke to him again, dead though he was. 'Patroclus,' he said, 'why be so sure of an early end for me? Who knows? Achilles, son of Thetis of the Lovely Locks, may yet forestall me by ending *his* life with a blow from my spear.'
Hector put his foot on Patroclus to withdraw his bronze spear from the wound, and thrust at the corpse till it came off the spear and fell face upwards on the ground.

Il. 17.192–208
When Zeus the Cloud–compeller saw Hector from afar equipping himself in the arms of the divine Achilles, he shook his head and said to himself: 'Unhappy man! Little knowing how close you are to death, you are putting on the imperishable armour of a mighty man of war, before whom all others quail. And it was you that killed

his comrade, the brave and lovable Patroclus, and stripped the armour from his head and shoulders with irreverent hands. Well, for the moment great power shall be yours. But you must pay for it.

Aen. 12.930–52
He, in lowly suppliance, uplifting eyes and pleading hands: "Yea, I have earned it," he cries, "and I ask not mercy; use thou thy chance. If any thought of a parent's grief can touch thee, I pray thee – in Anchises thou, too, hadst such a father – pity Daunus' old age, and give back me or, if so thou please, my lifeless body, to my kin. Victor thou art; and as vanquished, have the Ausonians seen me stretch forth my hands: Lavinia is thine for wife; press not thy hatred further."
Fierce in his arms, Aeneas stood with rolling eyes, and stayed his hand; and now more and more, as he paused, these words began to sway him, when lo! high on the shoulders was seen the luckless baldric, and there flashed the belt with its well–known studs – belt of young Pallas, whom Turnus had smitten and stretched vanquished on earth, and now wore on his shoulders his foeman's fatal badge. The other, soon as his eyes drank in the trophy, that memorial of cruel grief, fired with fury and terrible in his wrath: "Art thou, thou clad in my loved one's spoils, to be snatched hence from my hands? 'Tis Pallas, Pallas who with this stroke sacrifices thee, and takes atonement of thy guilty blood!" So saying, full in his breast he buries the sword with fiery zeal. But the other's limbs grew slack and chill, and with a moan life passed indignant to the Shades below.

Il. 22.326–71
As Hector charged him, Prince Achilles drove at this spot with his lance; and the point went right through the tender flesh of Hector's neck, though the heavy bronze head did not cut his windpipe, and left him able to address his conqueror. Hector came down in the dust and the great Achilles triumphed over him. 'Hector,' he said, 'no doubt you fancied as you stripped Patroclus that you would be safe. You never thought of me: I was too far away. Your were a fool. Down by the hollow ships there was a man far better than Patroclus in reserve, the man who has brought you low. So now the dogs and birds of prey are going to maul and mangle you, while we Achaeans hold Patroclus' funeral.'
'I beseech you,' said Hector of the glittering helmet in a failing voice, 'by your knees, by your own life and by your parents, not to throw my body to the dogs at the Achaean ships, but to take a ransom for me. My father and my lady mother will pay you bronze and gold in plenty. Give up my body to be taken home, so that the Trojans and their wives may honour me in death with the ritual of fire.'
The swift Achilles scowled at him. 'You cur,' he said, 'don't talk to me of knees or name my parents in your prayers. I only wish that I could summon up the appetite to carve and eat you raw myself, for what you have done to me. But this at least is certain,

that nobody is going to keep the dogs from you, not even if the Trojans bring here and weigh out a ransom ten or twenty times your worth, and promise more besides; not if Dardanian Priam tells them to pay your weight in gold – not even so shall your lady mother lay you on a bier to mourn the son she bore, but the dogs and birds of prey shall eat you up.'

Hector of the flashing helmet spoke to him once more at the point of death. 'How well I know you and can read your mind!' he said. 'Your heart is hard as iron – I have been wasting my breath. Nevertheless, pause before you act, in case the angry gods remember how you treated me, when your turn comes and you are brought down at the Scaean Gate in all your glory by Paris and Apollo.'

Death cut Hector short and his disembodied soul took wing for the House of Hades, bewailing its lot and the youth and manhood that it left.

Shades below.

SELECT BIBLIOGRAPHY

Fr. Cairns, *Virgil's Augustan Epic* (Cambridge 1989).
S. Farron, 'The death of Turnus viewed in the perspective of its historical background', *Acta Classica* 24 (1981), 97–106.
id., 'The abruptness of the end of the *Aeneid, Acta Classica* 25 (1982), 136–41.
id, 'Aeneas' revenge for Pallas as a criticism of Aeneas', *Acta Classica* 29 (1986), 69–83.
K.W . Gransden, *Vergil's Iliad* (Cambridge 1984).
J. Griffin, 'Achilles kills Hector', *Lampas* 23 (1990), 353–69.
S.J. Harrison, *Vergil. Aeneid 10, with Introduction, Translation and Commentary* (Oxford 1991).
Ph. Hardie, *The Epic Successors of Vergil; a Study in the Dynamics of a Tradition* (Cambridge 1993).
G.N. Knauer, *Die Aeneis und Homer* (Göttingen 1964).
id., 'Vergil and Homer', in: *ANRW* II 31.2 (1981), 870–918.
R.O.A.M. Lyne, 'Vergil & the Politics of War', in: S.J. Harrison (ed.), *Oxford Readings in Vergils Aeneid* (Oxford 1990), 316–38 (= *CQ* 33 [1983], 188–203).
id., Further Voices in Vergil's Aeneid (Oxford 1987).
L.A. MacKay, 'Achilles as model for Aeneas', *TAPhA* 88 (1957), 11–66.
V. Pöschl, *Die Dichtkunst Vergils. Bild und Symbol in der Aeneis*, 3rd. ed. (Berlin/New York 1977).
A. Powell, 'The *Aeneid* and the embarrassments of Augustus', in: A. Powell (ed.) *Roman Poetry and Propaganda in the Age of Augustus* (Londen 1992), 147–174.
M.C.J. Putnam, *The Poetry of the Aeneid* (Cambridge, Mass. 1965).
P. Schenk, *Die Gestalt des Turnus in Vergils Aeneis* (Königstein 1984).
P.H. Schrijvers, 'La valeur de la pitié chez Virgile (dans l' Énéide) et chez quelques–uns de ses interprètes', in: R. Chevalier (ed.), *Présence de Virgile* (Paris 1978), 483–95.
H.-P. Stahl, 'The Death of Turnus: Augustan Vergil and the political Revival', in: K.A. Raaflaub & M. Toher (edd.) *Between Republic and Empire. Interpretations of Augustus and his Principate* (Berkeley and Los Angeles, California 1990), 174–212.
D. West, 'The Deaths of Hector and Turnus', in: Ian McAuslan & Peter Walcott, *Vergil*, Greece and Rome Studies (Oxford 1990), 14–24.
G. Williams, *Technique and Ideas in the Aeneid* (New Haven 1983).

‘Death, the Elusive Thief’: The Classical Arabic Elegy

Pieter Smoor

This study deals with the Arabic elegy from the sixth to the tenth century, that is, from the pre-Islamic and classical period to the period of the New School, *al-Muḥdathūn*.[1] I will trace two developments, first, the impact or rather lack of impact of the Islamic notions of the Afterlife on the essentially irreligious outlook of the pre-Islamic period and, secondly, the most common motifs in elegiac poetry. One can indeed notice a certain development in the motifs of the genre from the pre-Islamic period until the first Islamic century, and then in the New School, the so-called *Muḥdathūn*, and there may also be traces of Hellenistic influence which are discernible.

There are, of course, many different ways in which human life can meet its end, and from the point of view of the elegy that was a good thing, for variety in the one led directly to variety in the other. The situation of a man who witnessed the death of seven sons in one day due to the fall of a boulder is very different from a woman's death caused when the camel she was riding started at the unexpected flight of a bird. The premature death of a child is of a very different order than the death of an older person struck down by a bolt of lightning. And what should one think of the mass grave at Badr into which the Prophet's fallen enemies were all thrown? Was that also worthy of an elegy in the eyes of a good Muslim?

1 More than one author has written extensively on the subject of the Arabic elegy: Ignaz Goldziher, “Bemerkungen zur arabischen Trauerpoesie” in *Wiener Zeitschrift für die Kunde des Morgenlandes* XVI (1902) (= I. Goldziher, *Gesammelte Schriften*, 6 vols., (Hildesheim 1967-’73), 4 : 361-393); and Ewald Wagner, *Grundzüge der klassischen arabischen dichtung*, 2 vols., (Darmstadt 1987), in particular 1 : 116-134 (= Chapter “VIII. Trauergedichte”), both of whom provide general descriptions of the genre, are examples. Other studies have been dedicated to the work of particular poets, such as the recent articles by Michael Winter “Content and Form in the Elegies of al Mutanabbī” in *Studia Orientalia D.H. Baneth Dedicata*, (Jerusalem 1979); and Derek Latham “The Elegy on the Death of Abu Shujāʿ Fātik by al-Mutanabbī” in *Arabicus Felix: Luminosus Brittanicus, Essays in Honour of A.F.L. Beeston on his Eightieth Birthday*, ed. Alan Jones, (Ithaca Press, Oxford University 1991). Arie Schippers has kindly shown me his article “Abū Tammām's elegies on Hālid ibn Yazīd al-Shaybānī” in the *Festschrift Ewald Wagner*, forthcoming. Schippers' study deals with the elegies which Abū Tammām composed on the death of a certain Khālid al-Shaybānī. These are poems which I will not deal with here, as the reader will soon see them treated in the aforesaid *Festschrift*, which by the way will also include a contribution of my own on the subject of a delirious sword: “The Delirious Sword of Maʿarrī: An Annotated Translation of his *Luzūmiyya Nūniyya* in the Rhyme-Form ‘*Nun Maksura Mushddada*’”.

Then there is the more genre-oriented critical literature which the Arabs themselves produced in the past. The 'Art of Poetry' by Ibn Rashīq (died in 456/1063-64, or in 463/1070-71) in particular stands out because of its direct style and clear opinions, while the material that was dictated later by Abu 'Alī al-Qālī (died 356/967) also provides interesting sidelights on the art of the elegiac.[2]

The story of the man who lost seven sons due to the fall of a boulder may not be historical, but we nevertheless possess a curious elegy on this subject which has survived. The first line of this elegy even demonstrates the repetition of words so common to the genre. We find this repetition in other early elegies, where it is spread out even further, over more than one line. The story of the Boulder and the Man who lost Seven Sons goes as follows:

> A man of the Banū Ḍabbah, in the time of the Heathens, had seven sons. These went out hunting with dogs. Later they sought refuge in a cave, but then a boulder fell on them and thus an end was made of their lives. When their father saw that there was no word of them, he went out to find them and followed their tracks to the cave. There he could find no continuation of their footprints. Then he knew that something terrible had happened, and he began to recite the following lines (a *Mīm* poem of which I cite only the following three verses):
>
> أسبعةُ أطوادٍ أسبعةُ أبحُرٍ   أسبعةُ آسادٍ أسبعةُ أنجُمِ
> رُزِئتُهُمُ في ساعةٍ جرّعتْهُمُ   كؤوس المنايا تحت صَخْرٍ مُرَضَّمِ
> فمَنْ تكُ أيّامُ الزمان حميدةً   لديهِ فإنّي قد تعَرَّقْنَ أعْظُمي
>
> 'Oh seven mountains, oh seven seas, oh seven lions, oh seven stars!
> A catastrophe has deprived me of them in only an hour, during which time they were forced to drink from the beaker of Death's Necessity while under a pile of fallen boulders.
> Whoever may enjoy the praises of Time's Days, know that with respect to myself they have scraped off the flesh of my bones'.[3]

This poem may not be genuine precisely because the repetition of words is limited to its first lines. Other elegies from the Heathen period are not so limited. Repetition, not of a few words, but of complete half-lines can be seen, for example, in an elegy that is contained in the Biography of the Prophet by Ibn Hishām. This elegy deals with those who fell at the Battle of Badr, when the bodies of the Prophet's opponents were all thrown down a well. These lines could indeed be

---

2 Ibn Rashīq, *Al-ʿumda fī maḥāsin al-shiʿr wa-ādābih wa-naqdih*, ed. M.M. ʿAbd al-Ḥamīd, 4th printing, (Beirut 1972);
Abu ʿAlī al-Qālī al-Baghdādī, *Kitāb al-amālī*, 2 vols., ed. Dār al-Kutub, Cairo, reprint (Beirut no date).

3 Abū ʿAlī al-Qālī, *Al-amālī*, 1 : 61.

authentic, although the elegist in question was probably not happy with the new religion of Islam:

تُحَيِّى بالسلامة أمّ بكْر   وهل لي بعد قومي مِنْ سلامِ
فماذا بالقليبِ قليبِ بدْر   من القَينات والشَرْب الكِرام
فماذا بالقليبِ قليبِ بدْر   من الشيِزَى تُكَلَّلُ بالسنام
وكم لكِ بالطَوِيّ طَوِيّ بدْر   مِن الحَوْمات والنَعَم المُسام
وكم لكِ بالطَوِيّ طَوِيّ بدْر   من الغايات والدُسُع العِظام

> Ummu Bakr gave me the greeting of peace;
> But what peace can I have now my people are no more?
> In the pit, the pit of Badr,
> What singing girls and noble boon companions!
> In the pit, the pit of Badr,
> What platters piled high with choicest camel-meat!
> In the well, the well of Badr,
> How many camels straying freely were yours!
> In the well, the well of Badr,
> How many flags and sumptuous gifts![4]

Here we see repetition in a number of different lines. This phenomenon can, it seems, be explained with reference to an even earlier phase in the development of the elegy, or *marthiyah*, as a distinct genre. It is thought to have grown out of the traditional wailing or *niyāḥah* which accompanied the deceased to the grave. The wailing, which was performed by women standing around the corpse, did rhyme after an irregular fashion, but was not metrical. Later, a simple meter was used, which became more complex as the regular *marthiyah* developed.[5]

We shall pause just for a moment by the rituals surrounding death in the pre-Islamic period and ask ourselves if there was then a belief in the immortality of the soul or a belief in the existence of a Paradise or Hell? It does not seem to be the case that such beliefs were held during this early time. One scholar has observed with some regret that the early Arab poets never felt a metaphysical dizziness when considering the phenomenon of death.[6] Indeed, reading the early Arabic elegies we seem to feel the presence of an all-powerful *Dahr*, a Fate of

---

4 *Al-Sīra al-nabawiyya li-Ibn Hishām*, ed. Muṣṭafā al-Saqqā' e.a., 2 vols., 2nd printing, (Cairo 1955 / 1375), 2: 29; and the translation in *The Life of Muhammad, a translation and notes* by A. Guillaume, (Pakistan Branch Oxford University Press, Lahore Karachi Dacca 1967), 352-53. The poem was composed by Abū Bakr ibn al-Aswad ibn Shuʿūb al-Laythī, 'Shaddād'.

5 See Goldziher, "Bemerkungen zur arabischen Trauerpoesie", 365-370.

6 Mohamed Abdesselem, *Le thème de la mort dans la poésie arabe des origines à la fin du IIIe / IXe siècle*, (Publications de l'Université de Tunis 1977), 112: , "les poètes ... n' ont éprouvé, au spectacle de la mort, nul vertige métaphysique."

Time or Destiny, which rules every aspect of human existence.[7] It seems possible to remove one of the veils covering the pre-Islamic period in order to learn something more about the *ghayb*, the 'Hidden Future.' Using the services of a fortune-teller (*kāhin*), the pre-Islamic Arab could try to discover what life had in store for him. Some fortune-tellers even claimed that they could determine when and where a person would die. In the following story such a divination plays a central role.

> A man named Ufnūn met a *kāhin* during the time of the Heathens. The *kāhin* said to him: 'You shall die at a place called Ilāha.' The man remained for some time where he was (so long as it was God's pleasure). Then he traveled in a caravan belonging to his clan that was going to Syria. They arrived at their destination, but when they departed from there they soon became lost. They asked someone who knew the area, 'In which direction should we travel?' The answer was: 'Continue on until you arrive at such and such a place where the road is clearly visible, then you will see Ilāha.' Ilāha is an elevated area in the steppe of Samāwa. When they arrived there, his companions got off their camels but Ufnūn refused to do so. His camel grazed while he remained in the saddle. Then a snake clamped itself fast into the camel's lip. It turned its head and rubbed its lip against Ufnūn's leg, which the snake then bit. Ufnūn said to his brother, Muʿāwiya, who was with him, 'Dig a grave for me because I'm going to die now.' Then Ufnūn sang a song before he died, in order to bewail his own death.

I shall only quote a part of this poem, which is no usual elegy but only a gathering together of lines that in an elegiac style refer to the imminent death of this particular poet himself:

فلستُ على شيءٍ فَرُوحَنْ معاويا ولا المُشفِقاتُ إذ تبعْنَ الحوَازيا
ولا خيرَ فيما يكذِب المرءُ نفسَه وتقوالِه للشيء يا ليت ذا ليا (...)
فطأ مُعْرِضاً إنّ الحُتُوفَ كثيرةٌ وإنّك لا تبْقي بنفسك باقيا
لَعَمرُكَ ما يدري امرؤ كيف يتّقي إذا هو لم يَجْعل له اللّه واقيا
كفى حَزنا أن يرحل الركبُ غدْوة وأنزَلُ في أعْلى الإلاهة ثاويا

> I can do no more, thus you must leave this place, o Muʿāwiya! Just as little as the pitying women can do, even when they consult the *ḥāziyah* s (the prophetesses). There is nothing good about deceiving a man's soul, also not when he says 'O, but I wish that was for me!' (. . .).
> Thus leave this place, the ways of meeting death are many, and you with your soul will also cease to exist.
> [By your life, a man does not know how to look after himself, when he fails to make God his Protector.]

7 Compare Werner Caskel, *Das Schicksal in der Altarabischen Poesie,* (Leipzig 1926).

> It is all too sad that tomorrow the caravan will continue on its way, whilst I must be left behind on the heights of Ilāhah.
> – Thus he died and they buried him.[8]

Death indeed appears in many guises, and this is also a topic treated in elegies. The moment of death is always determined by Fate. In cases where death has been caused by the agency of external violence, during a war, for example, the duty of vengeance must be fulfilled, the *tha'r*. The soul of the deceased can, according to the old thinking of the pre-Islamic period, take on the form of an owl (*hāmah*, *būm*, or *ṣadā*). The last word, *ṣadā*, also means 'thirst' and 'echo'. This bird knows no rest and is always to be found in the neighborhood of the grave. Whoever listens carefully can hear how this bird calls *isqūnī*, *isqūnī*, 'Give me something to drink, Give me something to drink!' The bird appears to thirst for the blood of the deceased's murderer.

Just such a bird also appears in a love poem, where the poet addresses a certain Laylā. When Laylā, his beloved, is unkind to him, he will die, and then she will be haunted by the graveyard owl. The bird will thirst after her blood, *Isqūnī*, Give me something to drink! For she is responsible for the murder of her despairing lover, who has pined away into death on account of her hard-heartedness. We have a poem on this theme in which an anonymous poet addresses a vague Supreme Being on the subject of his unhappy affair with Laylā:

وقال آخر :

فيا رَبٌ إنْ أهلكْ ولم ترو ِ هامتي بليلى أمتْ لا قبرَ أعطشُ من قبري

وإن أكُ عن ليلى سلوتُ فإنّما تسليتُ عن يأسٍ ولم أسلُ من صبر

وإن يكُ عن ليلى غنى وتجلّدٌ فربّ غنى نفسٍ قريبٍ من الفقر

8 The translation is a new one. The poem is the Mufaḍḍaliyya No. 65, and Ufnūn's name in full reads: Ṣuraym ibn Maʿshar ibn Dhuhl ibn Taym ibn Mālik Ḥubayb ibn ʿAmr ibn Ghanm ibn Taghlib. See *The Mufaḍḍalīyāt: an anthology of ancient arabian odes, compiled by al-Mufaḍḍal son of Muḥammad according to the recension and with the commentary of al-Anbārī*, ed. Ch. J. Lyall, vol. II, Translation and notes, (Oxford 1918), 202. The text describing the circumstances surrounding the poem reads slightly different in another ancient version, the *Sharḥ ikhtiyār al-Mufaḍḍal, ṣanʿat al-Khaṭīb al Tibrīzī*, ed. Fakhr al Dīn Qabāwah, (Damascus 1971-'72), No. 65, 1154ff..

The version in the Damascus edition is as follows:

"He rode on an ass. Then he refused to dismount with his companions; and after a long time passed, the ass suddenly lay down. At that moment he was bitten by a snake. They said: 'His ass is bitten and he is fallen!' He said to his companions: 'I am going to die.' They said: 'May no harm come to you' (or: 'Nothing is the matter.'). He said: 'Why then has the ass lain down?' This last line Ufnūn gives as a proverbial tag to allude to anything which itself is a proof that a situation happens to be totally different from the appraisal of it which others have communicated to you." (see *Sharḥ ikhtiyār al-Mufaḍḍal*, 1155).

Compare the same story later in Abū ʿUmar Aḥmad ibn Muḥammad Ibn ʿAbd Rabbih al-Andalusī, *Kitāb al-ʿiqd al-farīd*, ed. Aḥmad Amīn e.a., 2nd printing, (Cairo 1952), 3 : 247-248.

> Oh my Lord, if I die – without your having made the graveyard owl drink from Laylā's blood – then I will truly die indeed. No grave is thirstier than my grave.
> And if I forget everything, without having my Laylā, then shall I seek distraction driven by despair, I shall not forget owing to perseverance.
> If I become rich in my tenaciousness, without Laylā, then are many nearby soul-kingdoms stricken by poverty.[9]

The anonymous poet who was so much in love with one Laylā probably is Tawba ibn al-Ḥumayyir, her cousin. This poet also spoke of the graveyard owl when he expressed the wish that Laylā should come and greet him. He hoped that she would visit his grave while he lay inside it! He did indeed die, not from love, though, but in war, and because his soul remained unavenged, it appeared as a graveyard owl to claim its vengeance.

When Tawba invited Laylā to visit him, he promised to acknowledge her salutation, even if this should come to pass by means of the graveyard owl. The poet says:

ولو أنّ ليلى الأخيلية سلّمتْ　　　عليّ ودوني تربة وصفائحُ
لسلّمتُ تسليمَ البشاشة أو زقا　　　إليها صدىً من داخل القبر صائحُ
وأغبَطُ من ليلى بما لا أنالُهُ　　　ألا كلّ ما قرّتْ به العينُ صالحُ

> May it come to pass that Laylā al-Akhyaliyya should come to visit me when I am covered by dust and tombstone.
> Then I should, for my part, greet her with good cheer, or otherwise as graveyard owl (*ṣadā*) scream forthwith from out my grave.
> I am envious of Laylā, envious of something that I never even received. But I suppose that everything wherein the eye finds its consolation is good.

Laylā came later, much later to the grave, since she happened to be in its vicinity (she was actually married all of this time with someone else, but Tawba still found her charming). She called toward the grave, 'Peace be unto you, oh Tawba!' But no answer came! Disappointed, she addressed her companions: 'I never knew until now what a terrific liar he was!'
She was of course referring to the graveyard owl, which did not respond to her call – was it in the grave at all? In reality it was not, but there did happen to be an owl hidden in the shadow of the tomb. This owl was startled by the appearance of both Laylā's howdah or camel saddle, and the large number of her traveling companions. In panic it rose, flying up from its hiding place, and struck her camel in

---

9 My translation. *Sharḥ Dīwān al-Ḥamāsa* li-Abī 'Alī Aḥmad ibn Muḥammad ibn al-Ḥasan al-Marzuqī, ed. Aḥmad Amīn and 'Abd al-Salām Hārūn, 3 vols., 2nd printing, (Cairo 1967-'68), 3 : 1224-25. Cf. the translation in *Translations of Ancient Arabian Poetry, chiefly pre-Islamic, with an introduction and notes* by Ch.J. Lyall, (London 1930), 67.

its face. The camel jumped, and Laylā, no longer very young herself, had an unlucky fall which killed her. She was buried next to the poet she had come to visit. ...[10]

What is important here? There was first of all a *niyāḥa*, with repeated phrases of certain elements in the poem. In the time of the Heathens, before Islam, when no clear ideas of a life after death existed, there was still a belief that something of the deceased could continue to live on in the form of a graveyard owl. However, we still know very little about the structure which later forms of the elegy developed.

In the transitional period from paganism to Islam, during Mohammed's lifetime, there lived one of the best-known elegiac poets: a woman called al-Khansā', the Flat-Nosed One (because she was said to resemble a deer). She had a brother, Ṣakhr, who was to become as famous as she was on account of the elegies she composed for him. Another brother, for whom she did not lament so much, is for that reason much less well-known. We will deal with Ṣakhr instead, who once left on a raid against Bedouin of another tribe. In the course of the raid, he became badly wounded and as a consequence was seriously ill for weeks. Ṣakhr's wife was desperate. Whoever asked her how he was getting on received the following answer: 'He is not so alive as to give any reason for hope, nor so dead as to publish his obituary.' Ṣakhr's mother was apparently more optimistic and expressed herself quite differently, 'By virtue of Allah's good offices may he regain his health'.[11]

We understand from the sources that Ṣakhr was not very satisfied with his wife's attitude, for when he began to make a temporary recovery he decided to punish her severely. He tied her up to a post and was content to let her die there slowly. After she did die, he himself took a turn for the worse, and soon joined her.[12] Al-Khansā' was so impressed by what had happened, that she devoted the majority of her elegies to him. Whenever she could, she recited her elegies aloud with great clamor and much weeping. Her poetry, though one-sided, was certainly appreciated by her contemporaries. Only a single individual, the well-known satirical poet Jarīr, had to admit her superiority with a certain sullenness. When he was asked the usual critical question, 'Who is the best poet?' he answered: 'I am, were it not for the existence of that whore' (He meant al-Khansā').[13]

She is said to have lived a very long life, until far into the Islamic period. Her entire life she continued to lament over her dead brother. During the pilgrimage season she walked around in Medina while she lamented and wept. She was

---

10 *Sharḥ Dīwān al-Ḥamāsa*, 3 : 1311-12. Cf. *Translations of Ancient Arabian Poetry*, 76-77.

11 Abu l-Faraj al-Iṣbahānī ʿAlī ibn al Ḥusayn, *Kitāb al-aghānī*, ed. reprint Dār al-Kutub, (Cairo no date), 15 : 78 (only quotes Ṣakhr's wife.). Ibn Qutayba, *Kitāb al-shiʿr wa l-shuʿarā'*; (Leiden 1904), 198-199 (Ṣakhr's wife as well as his mother are quoted.)

12 For Ṣakhr meting out "punishment" to his wife, see Ibn Qutayba, *loc.cit.*

13 *Wa-qīla li-Jarīr: Man ashʿaru l-nāsi? Qāla: Anā lawlā hādhihi l-ʿāhira!'*, in *Shurūḥ Saqṭ al-zand* ed. Ṭaha Ḥusayn e.a., (Cairo 1945-'48), 4 : 1464-65.

dressed in a heathen outfit, although on her way to do the pilgrimage at Mecca. The caliph ʿUmar received many complaints about her behavior and resolved to speak to her about it: 'What you do is not part of Islam, and those you grieve for died in the Heathen period. They are sawdust for Hell.' Al-Khansāʾ answered, 'All the more reason for my grief.' Upon which she recited some more poems on the deaths of her brothers. The Caliph was impressed and told his fellow Moslems to 'Leave her alone, for she shall always be stricken by this grief.'[14]

From the stories about al-Khansāʾ, it appears that her poetry, despite the exclusive nature of her interest in writing only elegiac poems, nevertheless made a great impression on her contemporaries. Actually, she was not the only woman to excel in the composition of elegiac poetry, for there were others who were also well-known for the superb elegies they composed.[15]

We may consult the critic Ibn Rashīq for an explanation as to why the female poets were such talented elegists. Ibn Rashīq, who devotes a whole chapter to the elegy in his 'art of poetry', offers his opinion that women suffer painful emotions so much more quickly than men, yes, that they are 'the most feeling creatures when tragedy strikes, in all matters of the heart.' They are also 'the most tortured creatures with respect to the loss of a beloved one.' – Allāh is the sole responsible agent for how women are constituted, 'for He, the all-powerful One, gave to women that powerlessness and weakness of decision' that so clearly marks them off from men. This happens to be a good thing for literature, he remarks, because 'elegies are only composed on the basis of deeply felt pain'.[16]

Another reason of course is that in Arab culture the practical work of lamenting the dead is commonly women's work. The lament, or *niyāḥa*, is even named as such in the elegies of al-Khansāʾ, and this itself confirms the natural relation that exists between the elegy and the traditional lament.[17]

Generally speaking, the development of the elegy as a discernible genre begins to become manifest in the work of al-Khansāʾ. This kind of poetry, of the

---

14 N. Rhodokanakis, "Al-Hansâʾ und ihre Trauerlieder, Ein literar-historischer Essay mit textkritischen Exkursen", in *Sitzungsberichte der Philosophisch-historischen Klasse der Kaiserlichen Akademie der Wissenschaften* 147. Band (Vienna 1904), Abhandlung 4 : 8-9. Bint al Shāṭiʾ (ʿÂʾisha ʿAbd al-Raḥmān), *Al-Khansāʾ*, (Cairo 1963), 47. See in particular, Giuseppe Gabrieli, *I tempi, la vita e il canzoniere della poetessa araba al-Hansāʾ*, 2nd ed. by Francesco Gabrieli, (Rome 1944), 163-166.

15 Bint al-Shāṭiʾ, *Al-Khansāʾ*, 97. The text of quite a few dirges composed by women poets can be found inserted in an edition of al-Khansā's *Dīwān* by Louis Cheikho, *Anīs al-julasāʾ fī Dīwān al-Khansāʾ*, (Beirut 1889).

16 "*wa ʿalā shiddati l-jaza ʿi yubnā l-rithāʾu.*" See Ibn Rashīq, *Al-ʿumda*, 2 : 153.

17 Compare the following lines from an elegy by al-Khansāʾ, on the rhyme letter *ḥāʾ*:

(فنِسَاؤنا يندُبْنَ نو حاً بعد هاديةِ النوائِحْ)

"As if Time has cut our throats with butchers' knives, / Thus lamented our women with their keening, taking their cue from the leader of the dirge."
See Bint al Shāṭiʾ, *Al-Khansāʾ*, 101-102.

sort which she composed, presents a structure which would be more or less adopted 'whole cloth' by generations of later poets.

The beginning of an elegy by al-Khansāʾ is almost always an imperative address to the eye (of the poetess) to pour forth abundant and unhindered tears:

ألا يا عينُ ويحك أسعديني   لريب الدهر والدهر والزمن العضوض

> 'Oh eye woe unto you, help me confront the caprices of *Dahr* (fate) and the Sharp Tooth of Time.'

No tears may be spared, none, for example, for any catastrophe which may be yet to come:

ولا تبقي دموعا بعد صخر   فقد كلّفتِ دهرَك أن تفيضي

> 'Let no tears remain after Ṣakhr, for you [her eye] have been given the duty by Fate to inundate [her cheeks].'[18]

The second topic in al-Khansāʾ's elegies is that of the messenger who makes an announcement or presents a report concerning the death of the person whose loss is being lamented. Such an announcement, both in its conventional and natural forms, creates a terrible shock for the auditors. We hear al-Khansāʾ exclaim 'By my life!' and 'May you have no father!',

لقد صوّتَ الناعي بفقد أخي الندى   نداءً لعمري، لا أبا لك، يُسمَعُ

> 'The messenger who cries out that the 'brother with his lovely dew' [i.e., who was so generous] is missing, that messenger has come, with a report that – By my life! May you have no father! – clearly can be heard.'[19]

And elsewhere:

ألا ليت أمّي لم تلدْني سويّةً   وكنتُ تراباً بين أيدي القوابلِ
وخرّت على الأرض السماءُ فطبّقتْ   ومات جميعا كلّ حافٍ وناعلِ
غداة غدا ناعٍ لصخرٍ فراعني   وأورثني حزناً طويلَ البلابلِ
فقلتُ له ماذا تقول فقال لي   نعى ما ابن عمروٍ أثكلتْه هوابلي

> Oh me, would that my mother had not given me birth, would that I had been mere dust in the hands of the midwives.
> Would that the Heavens had fallen upon earth to wholly cover it up, that everyone – whether shod or unshod – had died in that calamity!

---

18 Bint al Shāṭiʾ, *Al-Khansāʾ*, 116.
19 Bint al Shāṭiʾ, *Al-Khansāʾ*, 104.

> On the morning that the messenger arrived with news of Ṣakhr's death, how he frightened me and left me the heir of a sorrow of long-lasting confusion.
> I said to him: 'What are you saying?' He answered me and gave the news that my mother had lost 'Ibn ʿAmr'.[20]

The death of Ṣakhr is responsible for some moments in al-Khansā''s poetry which are almost cosmic in effect (this is also a phenomenon in elegies which reappears in the work of later generations, where it is equally the case that the whole Cosmos is portrayed as suffering because of the passing of the beloved one). The example of al-Khansāʾ does not in any case make a too complicated impression on us:

يا عينُ جودي بالدمو — عِ على الفتى القرْم الأغرْ
والشمسُ كاسفة لمهــــــــــلكه وما اتّسق القمرْ
والإنسُ تبكي ولّهاً — والجنّ تُسعد من سمرْ
والوحش تبكي شجْوَها — لمّا أتى عنه الخبرْ

> Oh, eye, pour forth rich tears over this man, this radiant leader. (...)
> The sun is darkened by his loss, and the moon no more can shine as a full moon.
> The people lament with piercing grief, and the *jinn* s gave support to their complaint all through the night.
> Wild animals bewailed his loss in their pain when they heard of his death.[21]

The motif of both people and *jinn*s lamenting the loved one's passing is a recurrent one in elegies of later times, too. This can be seen in the work of a poet from the second century of Islam, namely Abu l-ʿAtāhiya. Ibn Rashīq explains in his 'Art of Poetry' why Abu l-ʿAtāhiya wanted to combine the laments of people and *jinn*s. He suggests that the method of the elegy must demonstrate that the survivors are clearly devastated by their loss, but it is possible that the emotions can show a mixture of sad longing as well as deep respect (*istiʿẓām*). This last element, the statement of respect, is appropriate whenever the dead person happens to have been a king or some other great leader. Abu l-ʿAtāhiyah shows this respect when lamenting the death of a caliph:

مات الخليفة أيها الثقلانِ — فكأنّني أفطرتُ في رمضانِ

'The caliph is dead, oh people and oh *jinn*s! (And it is as if I have broken the fasting in Ramaḍān).'

---

20 Bint al Shāṭiʾ, *Al-Khansāʾ*, 100.
21 Bint al Shāṭiʾ, *Al-Khansāʾ*, 101.

Ibn Rashīq gives the following comment in his 'Art of Poetry': 'When the standersby heard this verse, they raised their heads and opened their eyes wide. They said: He has announced the death of the caliph to the people and moreover to the *jinns*.' He adds that the magnificent effect of the beginning of this particular elegy is wholly destroyed by a really ugly half-line which immediately follows it.[22]

At any rate, it should be understood that the explicitly mentioned participation of *jinns* together with the ordinary humans in such a ceremony shows an extra value of respect toward the deceased person.

The poet Labīd who was converted to Islam during Mohammed's lifetime, and a contemporary of al-Khansā', gave a good example in his beautiful poems for the elegists of later generations. His elegies were highly valued and regarded as a model of inspiration. Labīd lost a brother due to an unexpected lightning bolt that struck forth from between the constellations of the stars and the rain clouds above. The brother, Arbad, must have been killed immediately. I would like to quote a few lines from the elegy which Labīd composed on the occasion of his brother's death. The fact that Arbad was killed by a lightning bolt is only indirectly suggested in the elegy. But just as in the elegies of al-Khansā', we have here, too, the repeated injunctions to the eye to shed all of its tears.

١) ما إن تعرّي المنونُ من أحدٍ   لا ولا والدٍ مشفقٍ ولا ولدِ

٢) أخشى على أربدَ الحتوفَ ولا   أرهبُ نوءَ السماكِ والأسدِ

٣) فجّعني الرعدُ والصواعقُ بالـــــــــــــــفارسِ يومَ الكريهة النجِدِ (...)

٩) يا عينِ هلاّ بكيتِ أربدَ إن   قمْنَ وقام الخصومُ في كَبَدِ

١٠) يا عينِ هلاّ بكيتِ أربدَ إنْ   ألوتْ رياحُ الشتاءِ بالعَضَدِ

1 The Fate of Death excuses no one, not the concerned father, not the son.
2 I feared different kinds of deaths would overtake Arbad, but I did not fear the rain-bringing stars of Simāk and the constellation Leo.
3 How painfully was I struck by the thunder and lightning which felled the rider on the day of terrible catastrophe. (...)
9 Oh my eye, have you not bewailed Arbad when not only they (the

---

22 Ibn Rashīq, *Al-ʿumda*, 2 :147-148, and *Abu l-ʿAtāhiya ashʿāruhu wa-ʾakhbāruh*, ed. Shukrī Faysal, (Damascus 1965 / 1384), 656-657.
Ibn Rashīq explains the second half of the line, "And it is as if I have broken the fasting in the month of Ramaḍān," as follows:
"Abu l-ʿAtāhiyah meant the following: — 'because I am making this statement public it is as if I am telling everyone that I had broken the fast during daytime in Ramaḍān. Everyone would then strongly disapprove of me and find that I had committed a great sin.'— (Ibn Rashīq then continues:) 'This is a good and an exceptional theme, but it is expressed in an ugly way; the poet was not able to tell what the soul really feels.'"

mourning women) but also the opponents began to set themselves against this catastrophe?
10 Oh my eye, have you not bewailed Arbad when the winter winds blew past dry sticks of wood? (...)[23]

The deceased Arbad is praised a little later in the poem because when he was still alive, he ensured that another party would have to mourn due to a murder he had committed:

١٢) حلوٌ كريمٌ وفي حلاوتهِ مرٌّ لطيفُ الأحشاءِ والكبدِ
١٣) ألباعثُ النوحَ في مأتمهِ مثلَ الظبا الأبكارِ بالجَرَدِ

12 '[Arbad was] sweet, noble, and in his sweetness, bitter, subtle in his liver and guts.
13 He caused the mourning of some other keening women – young gazelles in the steppes.'

Some standard motifs in later elegiac poetry can be traced back to another of Labīd's elegies on the death of his brother:

١) بلينا وما تبلى النجومُ الطوالع وتبقى الجبالُ بعدنا والمصانعُ
٢) وقد كنتُ في أكنافِ جار مضنّةٍ ففارقني جارٌ بأربدَ نافعُ
٣) فلا جَزِعٌ إن فرّقَ الدهرُ بيننا وكلّ فتاً يوماً به الدهر فاجعُ (...)
٥) وما الناسُ إلا كالديارِ وأهلها بها يومَ حلّوها وغدْواً بلاقعُ
٦) وما المرء إلا كالشهابِ وضوءِه يحور رماداً بعد إن هو ساطعُ (...)
١٥) فلا تبعدنْ إنّ المنيّةَ موعدٌ عليك فدانٍ للطلوعِ وطالعُ

1 We are worn out, but the rising stars do not decay, and mountains and castles shall remain after we are gone.
2 I was under the wings of a neighbor well-treated by all, but in the person of Arbad a good neighbor left me.
3 I am not a bending reed if Fate (*Dahr*) divides us twain. Each man will be painfully touched by Fate when the day comes. (...)
5 People are nothing more than places of residences and their inhabitants, but tomorrow the desert shall reclaim them.

---

23 *Der Diwan des Lebīd, nach einer Handschrift zum ersten Male herausgegeben* von Jûsuf Ḍijâ ad-Dîn al-Châlidî, (Vienna 1880), Poem V, 17-19. Cf. the German translation in *Die Gedichte des Lebîd nach der Wiener Ausgabe übersetzt und mit Anmerkungen versehn aus dem Nachlasse des Dr. A. Huber*, ed. Carl Brockelmann, (Leiden 1891), IX-X.

6 Man is nothing more than a flame which gives light: at first the flame is bright, but soon nothing but ash remains. (...)
15 Be not remote! (*Fa-lā tabʿadan* !) Death is a little piece of business which has been arranged for you, which is to begin soon or has already begun.[24]

This poem shows Man as a mortal creature when contrasted with stars, mountains and even castles, all of which are seen as immortal (or in any case of very long duration). The last line (15) contains a formula which later would become very frequently used and equally well known, namely that of *lā tabʿadan*, in the sense of 'Be not remote!' This sentiment seems rather odd since if anything is clear, it is that the deceased one must and will disappear into his grave. But it appears that the formula was pronounced in the hope that the deceased would provide protection to those who survived him.[25] Later this would become contradictory, for the motif in following generations would have the deceased still close to the survivors and at the same time infinitely far away, as far and unreachable as any of the constellations above.

Is it possible that the pre-Islamic Arabs believed that the dead lay in their graves in order to protect the town? It is certain that there was a custom in that period of leaving a camel (the *balīya*) by the grave of the deceased until it died.[26] There is also a story about three travelers, each of whom slaughtered a camel when passing a grave on their journey[27]. People also spilled water or wine on a grave, not only for the sake of the vengeful graveyard owl, but also in honor of the deceased.

Mohammed was not so enthusiastic about these practices, and he forbade in principle the custom of keening around the grave which was practised by mourning women. He is reported to have said that the dead one shall be punished if a professional mourner is hired to keen by the grave. Another tradition has it that this punishment is reserved only for those who during their lives gave express instructions for professional mourners to attend their funerals. Mohammed also forbade the transformation of graves into holy shrines.[28]

Another motif that was already present in one of Labīd's poems was the idea that mountains would always survive people, that they might even be immortal. Later, this motif returns in a somewhat different form. Someone is considered so important that he deserves immortal life. Such a person, then, is closer

---

24 *Der Diwan des Lebîd*, ed. al-Châlidî, (Poem VI), 21-24. Cf. the German translation in *Die Gedichte des Lebîd*, X-XII.

25 Mohamed Abdesselem, *Le thème de la mort*, 97.

26 Mohamed Abdesselem, *Le thème de la mort*, 101.

27 Abū ʿAlī al-Qālī al-Baghdādī, *Kitāb al-Amālī*, 2 : 143-145.

28 Compare Abu l-Qāsim Ḥusayn ibn Muḥammad "al-Rāghib al-Aṣbahānī", *Muḥāḍarāt al-udabāʾ wa-muḥāwarāt al-shuʿarāʾ*, 2 vols., (Beirut no date), 2 : 494 and 506 where we find the opinion of the Prophet on the professional female mourner: "The Prophet said: 'The professional mourner, if she feels no remorse before her own death, will arise on the Day of Judgement, but be held fast in skirts (*sirbāl*) of tar and an underskirt (*dirʿ*) of sulphur.'"

to being a mountain than a normal mortal. We find this motif already in the work of al-Nābighah al-Dhubyānī, who lived in the first century of Islam. He grieved for the death of an important clan leader named Ḥiṣn, in the following words:

يقولون حصنٌ ثم تأبَى نفوسهم   وكيف بحصنٍ والجبال جنوحُ
ولم تلفظِ الأرضُ القبورَ ولم يزلْ   نجومُ السماءِ والأديمُ صحيحُ
فعمّا قليل ثمّ جاشَ نعيُّه   فباتَ نديُّ القوم وهو ينوحُ

1 They say Ḥiṣn! . . . Then their souls refused to listen. How could this befall Ḥiṣn, while the mountains are still visible?
[the mountains are thus not split into two, Ḥiṣn is immortal, just like a mountain].
2 The earth has not vomited up its graves, the stars are still in the heavens, and the surface of the earth is undamaged.
[the cosmos is as it used to be, thus Ḥiṣn, as a cosmic entity, must still be living].
3 But shortly the swelling noise accompanying the news of his death breaks loose, then suddenly the mood of the clan in conclave changes to a mood of complete and undiluted lamentation.
[The news of his death is limited to what is said in the last line of this little poem.][29]

The 'topsy-turvy' theme of the mountain, the thought that 'a dead man can be so important that he is more like a mountain than a mere mortal' appears again a century later in the work of the poet and prince, Ibn al-Mu'tazz. This poet of the New School (the *Muḥdathūn*) composed a short epigrammatic elegy on the death of an important vizier, one Abu l-Qāsim:

وقال يرثي عبيد الله بن سليمان:
قد استوى الناسُ وزال الكمال   ونادتِ الأيّامُ أين الرجال
هذا أبو القاسمِ في نعشِهِ   قوموا انظروا كيف تزول الجبال
يا ناصرَ الملكِ بآرائه   بعدَك للملك ليالٍ طوال

He said in mourning the death of 'Ubayd Allāh [Abu l-Qāsim] ibn Sulaymān the following:
1 'People have become each other's equals, and perfection has ceased to exist. The days call, 'Where have the heroes gone?'

29 *Dīwān al-Nābigha al-Dhubyānī*, ed. Muḥammad Abu l-Faḍl Ibrāhīm, (Cairo 1977), 190: a dirge of only three lines dealing with Ḥiṣn ibn Ḥudhayfah al-Fazārī. This poem is quoted also by Ibn Rashīq, *Al-'umda*, 2 : 147.

> 2 This is Abu'l-Qāsim lying on his bier: Arise, and see how the mountains are leveled!
> 3 Oh, you who supported the Kingdom through your insight. After your death, the only thing that remains for the Kingdom are long nights.'[30]

Alongside the continually repeated motifs, which we find returning in various subtle forms in the poetry of different generations and centuries, there are other aspects which we should notice with respect to the structure of this kind of poem.

An important aspect in the elegy is the nature of the death being dealt with: a poet composing an elegy on the death of a woman or child had no easy task. Ibn Rashīq claims in his 'Art of Poetry' that these subjects are so extraordinarily difficult to elegize because the poet is constrained by the limited nature of possibilities open to him. Small children who have died have realized so little of their lives.[31] Women, who were always so well hidden from view and protected by jealous husbands or their own families, are also 'unknown' quantities whom it is difficult to describe.[32]

It had long ago been recognized that a detailed description of a woman could cause serious difficulties. The heathen poet al-Nābigha described the queen of Ḥīrah so realistically in a poem of praise that he had to flee the court and the outraged king for his life. The poet al-Mutanabbī, on the other hand, of the fourth Moslem century, was much more careful in his elegies. During his stay in Aleppo at the court of king Sayf al-Dawla, three noble ladies chanced to pass away. As court poet, he was required to compose the appropriate eulogies. Who were the ladies in question? All of them were members of the king's own family, namely his mother, his younger sister and finally his older sister. The three women died while living at one of the king's most remote fortresses in Mayyāfāriqīn in North Syria, where they had been removed for their own 'protection'.[33]

Because the women in question were of such high rank, the poet did not feel free to lament in a manner that might be construed as too intimate or moving.

---

30 *Dīwān ashʿār al-amīr Abi l-ʿAbbās ʿAbd Allāh Ibn Muḥammad al-Muʿtazz billāh al-Khalīfa al-ʿAbbāsī*, ed. Muḥammad Badīʿ Sharīf, (Cairo 1978), 2 : 358. This poem is also quoted by Ibn Rashīq, *Al-ʿumda*, 2 : 150 where the last half of line 2 reads differently:

قوموا انْظُرُوا كيف تسير الجبالُ ....

"... arise and see how the mountains journey forth!"

31 For elegies on children, see A. Schippers, "Abū Tammām's 'unofficial' elegies" in *Union Européenne des Arabisants et Islamisants, 10th Congress Edinburgh 9-16 September 1980, Proceedings*, ed. by Robert Hillenbrand, (Edinburgh 1982), 101-106.

32 Ibn Rashīq, *Al-ʿumda*, 2 : 154 and 158.

33 Nuʿm, the mother of Sayf al Dawlah, died in 337/948. His younger sister died in 344/955, while the older sister, "Sitt al-Nās" Khawla, survived until 352/963. The fortified city of Mayyāfāriqīn was not only the residence of these three women in life. After their deaths they were also buried there, because it happened to contain the necropolis of the Ḥamdānid dynasty. For more information on Mutanabbī's life and poetry see R. Blachère, *Une poète arabe du IVe siècle de l'Hégire (Xe siècle de J.-C.): Abou ṭ-Ṭayyib al-Motanabbî*, (Paris 1935).

Rather, he felt constrained in this very public situation to show reserve and respect. Still, he was criticized for this attitude, especially with regard to the lament he composed for the king's mother. The most devastating criticisms were leveled by his personal enemy, one al-Ṣāḥib Ibn ʿAbbād, an important vizier from the Persian capital Rayy (near present-day Teheran), who happened to be interested in literature.[34]

We do not need to agree with this particular criticism in order to interest ourselves in the larger question of how Arabic criticism manifested itself. At the same time I can indicate certain aspects of the elegy on the death of Sayf al-Dawla's mother which I find noteworthy. In the poem for the deceased mother of the King we see a number of ideas or, better, motifs to which al-Mutanabbī gives poetic expression. I will give a bird's-eye view of the motifs in question [for a full translation of the text in question, see the APPENDIX at the end of this article] :

First, there is the motif of the futility of all human action. 'We' make our swords and spears ready for battle, but Death strikes without weapons and with no battle necessary: no flight is possible, no matter how fast our thoroughbreds may be (lines 1-2). The World (*al-Dunyā*, a feminine word) is an attractive woman, who is unfortunately shut away from all suitors (she doesn't make dates with her lovers). Life itself is perhaps just as beloved as a fascinating woman, but in fact offers only a dream vision of that woman. The Fate of Time (*Dahr*) shot arrows toward 'me' (the poet's persona) and 'I' thus became a quiver full of arrows, each of which pierced the other. 'I' meanwhile understood that there was nothing at all I could do to ward off Fate (lines 3-7).

It is not until line 8 (following the poet's ominous introductory lines) that the first mention of the death in question is communicated. The news of the death which is the poem's subject is said to be of such an earth-shaking kind that the poet states that neither he nor any of his readers had ever heard news of such tremendous import before!

In line 10 a wish is suddenly expressed, which the poet utters in order to bring God's blessing down upon the deceased lady:

صلاةُ اللهِ خالقِنا حنوطٌ   على الوجْه المكفّنِ بالجمالِ

10 'May God's blessing, that of our Creator, be a balsam upon the face that is now covered in a winding-sheet of beauty.'

Mutanabbī's own contemporaries were critical of this particular line. The line itself might be beautiful, but it went too far in suggesting that the deceased mother of the king was also beautiful. In reality, one could hardly speak of beauty

---

34 For the critisism in question, see al-Ṣāḥib Abu l-Qāsim Ismāʿīl Ibn ʿAbbād (lived 326-385 H.), *Al-kashf ʿan masāwiʾ shiʿr al-Mutanabbī*, ed. al-Shaykh Muḥammad Ḥasan Âl Yāsīn, (Baghdad 1965), 45-48, where Ibn ʿAbbād criticizes the elegy written on the occasion of the mother's death.

when describing the physical appearance of a woman as old as Methuselah, as you will agree. According to Ibn Rashīq, Mutanabbī's contemporaries were said to have muttered, 'What connection did Mutanabbī have with this old woman that he felt compelled to describe her as beautiful?'

Ibn ʿAbbād, Mutanabbī's enemy, also had some criticism of his own to vent concerning this line: 'this is the kind of metaphor which a blacksmith would use at a wedding banquet.' The vizier was referring to the combination of the two words 'winding-sheet' and 'beauty,' but Ibn Rashīq also gave it as his opinion that there was nothing really exceptional in this, as Mutanabbī uses similar combinations elsewhere as well.[35]

I will not belabor this point but will turn now to the rest of the poem's content and its use of motifs. Line 11 shows us an antithesis between two kinds of veils: the buried lady during her life made use of a veil or protective-covering that was made of 'Good Morals', but now that she has passed on into the realms of Death, she has exchanged this veil for another no less strong, which is the layer of earth that covers her.

In the following lines the poet shares with us his reflections on the subject of the deceased, and this is followed in line 13 by the proverb, 'No one is immortal, and the World itself shall end.' Line 14 suddenly directs our attention toward a particular person (the stylistic figure *iltifāt*), for at this moment the addressee becomes identifiable as the deceased lady herself:

١٤) أطابَ النفسَ أنكَ متَّ موتاً   تمنّتهُ البواقي والخوالي
١٥) وزلتِ ولم تريْ يوماً كريهاً   تسَرَّ النفسُ فيه بالزوالِ

14 It was good for the soul that you have died in a way that all women wish for, both those who are already deceased and those who are their survivors!
15 You have left us, but you had never witnessed a horrible day, one so horrible that your soul would have been made happy by a journey (toward Eternal Life).

Thus the poet opines that the occasion of this death could not have come earlier. The lady would only have left this life had she been subjected to something horrible, but during her long life nothing of that kind had ever occurred. Line 16, which follows here, was unfortunately also the subject of criticism:

١٦) رواقُ العزِّ فوقكِ مسبطرٌّ   ومُلكُ عليِّ ابنكِ في كمالِ

16 'A gallery of pillars in the Hall of Fame stretches out above you (*fawqa-ki musbaṭirrun*) because the kingship of your son, ʿAlī, was one of perfection.'

35 Ibn Rashīq, *Al-ʿumda*, 2 : 154 and al-Ṣāḥib ibn ʿAbbād, *Al-kashf*, 47.

The criticism referred to is directed toward the phrase 'stretches out above you', which people thought to be too low in style considering the subject was the mother of a king.[36] But it is naturally very clear to us that the poet here is both lamenting the mother and flattering her son, the king. Perhaps ʿAlī, or Sayf al-Dawla, might be encouraged to reward the poet for his efforts with an even greater gift? In the following lines, 17-18, the poet expresses the wish that a heavy rain cloud shed its water on the grave, a not uncommon phenomenon in the older elegiac poetry. Lines 30-31 are interesting:

٣٠) ولا مَن في جنازتها تِجارٌ   يكون وداعُها نفضَ النعالِ
٣١) مشى الأمراءُ حوليْها حفاةً   كأنّ المروَ من زفِّ الرئالِ

30 There were no peddlers present at her burial ceremony, no one who took farewell of her by slapping the sand out of their sandals.
31 The emirs walked barefoot around her body, as if the white flint (around the grave) consisted of the fine feathers of ostrich chicks.

Here the poet indicates that the deceased lady was of high rank, for no one who walked around her grave was wearing sandals or slapped them afterwards to rid them of sand. No, the only people there are all emirs, and they are all barefoot out of respect for the deceased, but they pretend not to feel any pain as they walk on the sharp flints that surround the grave.

Lines 34 and 35 allow us to see how difficult it must have been for al-Mutanabbī to praise the old lady. He can't think of anything better to say than that she is superior to men, and as such exceptional for a woman:

٣٤) ولو كان النساءُ كمَن فقدْنا   لفُضّلتِ النساءُ على الرجالِ

34 'If all women were as the one we have lost, then all women would be preferred above men.'

The poet goes a step further with this motif. In Arabic the word for 'sun' (*shams*) is grammatically female, and this, he suggests, must be a mistake, it should naturally be male. The sun is, after all, an exceptional and radiant heavenly body, which gives off much more light than, for example, the moon (*qamar*, 'moon' and *hilāl*, 'waxing moon' are both grammatically male in Arabic). But considering the preference for women on account of the exceptional old lady, the poet now thinks differently about the word 'sun':

---

36 Ibn Rashīq, *Al-ʿumda*, 2 : 155 and al-Ṣāḥib ibn ʿAbbād, *Al-kashf*, 46. In F. Dieterici's edition of al-Mutanabbī's poems, *Dīwān al-Mutanabbī with the commentary by Abi-l-Hasan Ali ibn Ahmad al-Wahidi al-Naisaburi (died 468 A.H.)*, (Berlin 1861), 390-391, another word can be found for the rejected word *musbaṭirrun* ('stretches out') in the commentary on line 16: a servant of al-Mutanabbī is said to have coincidentally been passing by when some professors of language were criticizing this very passage. According to the servant, his master (al-Mutanabbī) had a different formulation of the line in question: "A gallery of pillars in the Hall of Fame is covered in shadow above you (*fawqa-ki mustaẓillun*)." In this way the line would have given less offense.

٣٥) وما التأنيث لاسم الشمسِ عيبٌ　　ولا التذكيرُ فخرٌ للهلالِ

35 ‘The attribution of the female gender to the word *shams* (sun) is no shameful matter. The attribution of the male gender to the *hilāl* (waxing moon) is no claim to fame!’

Lines 37 and 38 are impressive, and we find them again in different variations in the work of later poets. They read as follows:

٣٧) يدفّن بعضنا بعضاً وتمشي　　أواخرنا على هامِ الأوالي
٣٨) وكمْ عينٍ مُقبّلةِ النواحي　　كحيلٌ بالجنادل والرمالِ (...)
٤٠) أسيفَ الدولةِ استنجدْ بصبر　　...

37 ‘Some of us bury others of us; the last among us walk upon the skulls of the first.
38 Many an eye that once was kissed with all that was beautiful surrounding it is now made-up with a cosmetic (*kuḥl*) that consists only of pebbles and sand.’

Line 40, ‘Oh Sayf al-Dawla, look for support through perseverance!...,’ etc., concludes the elegy already in a certain sense, for with this line later generations would begin the very usual condolence-section of the elegy (*taʿziya*).

This section of the present elegy is one which we will not discuss now, for it is, if anything, in praise of the son and does not express an elegiac sentiment with respect to the deceased mother. The son, who is both next of kin and heir, is presumably invited to reach an accord with the poet as to the size of the reward that may be considered appropriate.

The unusual thing about this poem, which was composed some 300 years after the rise of Islam, has to do with the position it accords to Fate: Fate is certainly present in the poem, but there is no room left over for any kind of description of a Paradise. Perhaps it was felt that the pleasures of Paradise, which al-Mutanabbī does not touch upon, were reserved for a later time, pushed forward, as it were, to the Resurrection of the Dead and the Day of Judgment. In any case, al-Mutanabbī's poem allows us to see a good example of a Bedouin elegy of an archaic type. The human qualities of the lamented, highly placed lady are wholly covered in his poem by other qualities which are decorous stereotypes. A general remark with respect to the positive discrimination of the female sex, and that only because the deceased individual in question happened to be the king's mother, tells us in fact very little about her actual personality.

Conclusion: We have seen how elegies for women could be especially adapted. We have also seen how strong the role of Fate is in the motifs of the elegy. The role of Fate continued as a presence in these poems, despite the appearance of Islam, which teaches us that ‘Fate’ is nothing else than ‘God’ or ‘Allāh.’ The elegy and all its conventions originate from the pre-Islamic period and probably were only able to adapt themselves to the advent of Islam slowly and with difficulty.

Insofar as the influence of other cultures outside of Islam is concerned, we do now and again come across the name of one or another scholar, for example Galenus, but there is nothing systematic about such allusions.[37] One could suggest that the style-figures which would appear in later Arabic poetry imply a more rationalistic and philosophical approach to life. In any case, the activity of translators in Iraq and Syria certainly increased the knowledge of Hellenistic thought among readers of Arabic.

In this connection we may also bring to mind a brief elegy that was composed by the second century (Anno Hijra) Arabic poet, Abu l-ʿAtāhiya. This poet was of an ascetic cast of mind, having understood that in his own day this was the best way to increase his popularity among the common people and among those highly placed figures, such as Hārūn al-Rashīd, for whom piety was akin to intoxication. But I digress.

Insofar as the poem of Abu l-ʿAtāhiya is concerned, the last line suddenly confronts us with an allusion to material from the Alexander legend. I cite here only the last few lines of this short poem:

وقال يرثي عليّ بن ثابت الأنصاري:

٤) بكيتكَ يا عليُّ بدمعِ عيني   فما أغنىَ البكاءُ عليكَ شيّا

٥) كفى حَزَناً بدفنِكَ ثمّ إنّي   نفضتُ ترابَ قبْرك من يديّا

٦) وكانتْ في حياتِك لي عِظَاتٌ   وأنت اليومَ أوْعظُ منك حيّا

(He says the following, by way of lamenting ʿAlī ibn Thābit al-Anṣārī . . .)
4 I have mourned you, oh ʿAlī, with the tears of my eye, but all that lamenting has failed to do you any good.
5 It was sorrowful enough for me to have had to bury you, and then to have had to shake off from my hands the dust of your grave.
6 Your life already contained warnings in my direction, but today you warn me more than you ever did when still alive![38]

In the last line of the elegy, Abu l-ʿAtāhiya makes use of a motif from the Alexander legend, which also continued to live in the Islamic world. A well-

37 Wise Galenus is mentioned in the work of al-Mutanabbī. In an elegy on the deceased aunt of the Persian Grand Duke, the Būyid Sultan ʿAḍud al-Dawlah, the poet described Death as something that misses no one and levels all differences in the grave:

"The shepherd dies in his ignorance the same death as the learned Jālīnūs, whose mind knew all medical science."

(Line 15, from a poem on rhyme *qalbihi*, in *Dīwān al-Mutanabbī*, ed. Dieterici, 783; also in *Dīwān al-Mutanabbī*, ed. Dār Bayrūt li l-ṭibāʿa wa l-nashr, Beirut 1970, 558.)

38 *Abu l-ʿAtāhiya, ashʿāruh*, 675-679, text and notes to an elegy on rhyme *ladayyā*.

known Arabic prose writer from Baṣra (scil., al-Jāḥiz) knows, for example, that rhetoricians were said to have stood at the head of Alexander the Great's corpse. One of them is supposed to have spoken the following words, which later became so famous: 'Al-Iskandar was more eloquent yesterday than he is today, whereas today he warns us more severely than he did yesterday' (*Al-Iskandaru kāna amsi anṭaqa minhu l-yawma, wahuwa l-yawma awʿaẓu minhu amsi*).

The story was gratefully adopted by later generations of authors. They even dug up the name of the person who is said to have uttered these words while standing at Alexander's head. It was thought to be the famous Dayūjānus (Diogenes) who wisely said that Alexander 'was yesterday more eloquent than he is today.'[39]

Almost the same words are said to have been uttered, according to a certain Arab historian (al-Masʿūdī, *Murūj al-dhahab*), in the salon of the Caliph, al-Wāthiq, in Baghdad. At this occasion, Abu l-ʿAtāhiya's little poem was also recited as a version of the same motif. The Caliph, who was present at the salon, was much moved at hearing such wisdom spoken. The historian describes the deeply affected Caliph as follows: 'Al-Wāthiq wept terribly and his moans were loud, whilst all others present at the Salon kept his moans company with theirs.'[40]

Some centuries later again, another Arab prose writer (al-Rāghib al-Aṣbahānī, *Muḥāḍarāt al-udabāʾ*) mentions an alternative speaker of the Alexander proverb. The wise man who uttered the famous words was, according to him, none other than the well-known teacher of philosophy, Arisṭāṭālīs![41] Despite the disagreements in the Arab world as to which person may be justly credited with uttering the famous words, their effect on Arabic poetry is no more than an incidental one.

The Hellenistic element, especially within the special genre of the Elegy, does not seem very great. The changes that the elegy underwent in the first centuries of Islam were mostly internal in nature, namely those with respect to the ideas and motifs of the genre itself.

In the beginning, the elegy emphasized the importance to the tribe of the loss incurred by a death, and encouraged revenge. The members of the family were encouraged to lessen the thirst for blood felt by the graveyard owl in which the dead man's soul was now thought to have found its abode. Later, the elegist was able to place more emphasis on the individuality of the dead person. Fate remained responsible for the journey of the soul (*nafs*) or the spirit (*rūḥ*) of the deceased, with a less important role played by the tribe. As we have seen, the role played by Fate is still important in the work of a relatively late poet such as al-Mutanabbī. But in the work of other poets of the same period, such as Abū Tammām and al-Maʿarrī, there was indeed the idea of a compensation for earthly pain in the form of Paradisiac pleasure. More on this some other time, hopefully before the Resurrection of the Dead comes along!

---

39 *Abu l-ʿAtāhiya, ashʿāruh*, 675-677.

40 *Abu l-ʿAtāhiya, ashʿāruh*, 677.

41 *Abu l-ʿAtāhiya, ashʿāruh*, 678.

APPENDIX

Abu l-Ṭayyib al-Mutanabbī lamented Sayf al-Dawlah's mother and condoled him on the occasion of her death in 337 Hijrah [*Dīwān al-Mutanabbī, with the Commentary by Abi-l-Hasan Ali ibn Ahmad al-Wahidi al-Naisaburi (died 468 A.H.*), ed. F. Dieterici, Berlin 1861, 388-390; and ed. Dār Ṣādir, Beirut, 1970, 265 ff.].[42]

1 We make the Mashrafitic swords and spears ready, but Death's necessity destroys us just like that, without a battle!
2 We tie the thoroughbred horses fast near the tents, but still they do not allow us to escape the galloping night.
3 Who did not in years past love the World? But now there is no path that leads to a rendezvous!
4 The part of your life which you give to your lover is as the part of your life which is spent sleeping, phantom-like (i.e., love is as unattainable as the vision of a dream).
5 The Fate of Time has brought catastrophe upon me, my heart is encased in a quiver of arrows.
6 My situation is as follows: as soon as the arrows reach me, the new arrows split the old ones in two.
7 Then Destiny became unimportant to me, for the disasters which befell me could touch me no longer - worry and concern changed nothing.
8 — This is the first to bring news of a death, the first death to occur amongst so much majesty.
9 As if death (before this) never had caused anyone grief, as if the thought of death had never before occurred.
10 God's blessing, He who is our Creator, may it fall as balsam upon the face that is wrapped in a winding-sheet of beauty.
11 Protection for the buried one consists of Noble Traits, which were there even before the protection of the Dusty Grave.
12 Our thoughts of him, while he decays, provides the buried one in Earth's lap with a new birth (his good reputation now begins to live).
13 No one of all the creatures God has made can be eternal, no, the World would first move towards its last day.
14 It did good unto the soul that you died in a way that all women must envy, those already dead as well as those still living!
15 You have left us, but you had never experienced a horrible day, so horrible that your Soul would have been glad to depart (for the Life after Death).
16 A gallery of pillars in the Hall of Fame stretches out above you, for the Kingship of your son, ʿAlī, was one of perfection.

---

42 The translation is a new one. There is also a translation by A.J. Arberry, *Poems of al-Mutanabbī, a selection with introduction, translations and notes*, (Cambridge 1967), 56-62 (= Poem Nr. 10).

17 May it come to pass that your resting place enjoy the rain from the morning's clouds, which are as generous as your hand.
18 The rain which falls upon the tombstones pours down abundantly and is as the clashing hooves of horses which have seen the sickles cutting grain.
19 After your decease I will be suspicious of all Fame, for I cannot grow accustomed to Fame which does not have You as its subject.
20 One begging favors passes your grave and bursts forth weeping: his tears prevent him from asking the favor he came for.
21 How many benefits would you give him, if you were now free to dispose of your largesse?
22 By your life! Have you forgotten, perhaps? Although I am far from your land, my heart does not forget your generous gifts.
23 In despite of us you have descended to a place where you are far from the soft Zephyr and raw North wind.
24 The odor of perfumes which is Khuzāmā's does not reach you there, the rain is kept out.
25 In a place of residence (i.e., the grave) where each inhabitant is a stranger, alone for a long time and cut off from all ties of family.
26 Well-kept is she there, as water from the clouds: she closely guards all secrets and speaks with full integrity.
27 A doctor of medicine nursed her, while her son is a doctor of spears:
28 As soon as a sickness is described along one of his borders, he cures it with the points of his long spears.
29 She is not as the women for whom one gets the grave ready in order to serve as a veil (she was already modestly veiled in life).
30 No peddlers attended her burial ceremony, no people who slapped the sand from their sandals in farewell.
31 The emirs walked barefoot around her, as if the white flints (around the grave) were made of the fine feathers of ostrich chicks.
32 The pavilions displayed the beautifully formed women in the eye of the public, with ink put on where they usually were wont to apply expensive perfumes (they were made-up for mourning).
33 The disaster hit the women when they were thinking of no such thing: tears of sorrow flowed through those other, flirting tears.
34 If all women were as she whom we have just lost, then women would be preferred to men.
35 Applying the feminine gender to the noun *shams* (sun) is no shame. Applying the male gender to the *hilāl* (waxing moon) is no claim to fame!
36 The most lamented of those who pass away is one for whom no equal could be found up until her departure hence.
37 Some of us bury others; the last among us walk upon the skulls of the first.
38 Many an eye that was once kissed with everything beautiful around it, is now adorned with a make-up (*kuḥl*) that consists only of pebbles and sand.
39 How often has one closed his eyes, although he never closed his eyes before, for anything important. How often has one tended to complain of a loss of weight, when now he is shriveling in his grave.

40 Oh Sayf al-Dawla, find support through perseverance! How can one find as much perseverance in the mountains as you seem to possess? [even mountains do not possess as much]
41 You shall teach all men to find consolation, and how to wade through Death in a time of heavy war.
42 Time's circumstances are manifold with respect to yourself, but you are unique with respect to each of Time's moments.
43 May your sea never sink into dryness, Oh full sea! – Even though camels from far come to drink more than once, even though camels for the second time are drinking among those who could only drink once.
44 I believe that you, whom I see among kings, march ever straight forward while the world turns absurdly around you.
45 Thus, if you are superior to all others, although you are a part of mankind, you are like (the precious) musk: it, too, is only a part of the great stag's blood.

BIBLIOGRAPHY

ʿAbbās, Iḥsān, *Shiʿr al-Khawārij*, collection and edition, Beirut, 3rd printing 1974.
Abdesselem, Mohamed, *Le thème de la mort dans la poésie arabe des origines à la fin du IIIe / IXe siècle*, Publications de l'Université de Tunis, 1977.
Abu l-ʿAtāhiyah, ashʿāruhu wa-ʾakhbāruh, ed. Shukrī Fayṣal, Damascus 1965 / 1384.
Arberry, A.J., *Poems of al-Mutanabbī*, selected & edited, Cambridge 1967.
Bint al-Shāṭiʾ (= ʿÂʾisha ʿAbd al-Raḥmān), Al-Khansāʾ, Cairo 1963.
Blachère, R., *Un poète arabe du IVe siècle de l' Hégire (Xe siècle de J.-C.): Abou ṭ-Ṭayyib al-Motanabbî*, Paris 1935.
Bräunlich, E., "Versuch einer literaturgeschichtlichen betrachtungsweise altarabischer Poesien.", *Islam* XXIV (Leipzig 1937), pp. 201-269.
Caskel, Werner, *Das Schicksal in der Altarabischen Poesie*, Leipzig 1926.
Cheikho, Louis, *Anīs al-julasāʾ fī dīwān al-Khansāʾ*, iʿtanā bi-ḍabṭihi wa-tabwībihi aḥadu l-ābāʾ al-Yasuʿiyyīn (Louis Cheikho), wa-ḍamma ilayhi marāthiya sittīna shāʿirah min shawāʿiri l-ʿArab, Beirut 1888.
Der Diwan des Lebīd nach einer Handschrift zum ersten Male herausgegeben von Jûsuf Ḍijâ-ad-Dîn al-Châlidî, Wien 1880.
Dīwān ashʿār al-amīr Abi l-ʿAbbās ʿAbd Allāh Ibn Muḥammad al-Muʿtazz bi l-Lāh al-khalīfa al-ʿAbbāsī, ed. Muḥammad Badīʿ Sharīf, Cairo 1978.
Dīwān al-Mutanabbī, ed. Dār Bayrūt li l-ṭibāʿa wa l-nashr, Beirut 1970.
Dīwān al-Mutanabbī with the commentary by Abi-l-Hasan Ali ibn Ahmad al-Wahidi al-Naisaburi (died 468 A.H.), ed. F. Dieterici, Berlin 1861.
Dīwān al-Nābigha al-Dhubyānī, ed. Muḥammad Abu l-Faḍl Ibrāhīm, Cairo 1977.
Gabrieli, Giuseppe, *I tempi, la vita e il canzoniere della poetessa araba al-Hansāʾ*, 2nd ed. by Francesco Gabrieli, Rome 1944.
Die Gedichte des Lebîd, aus dem Nachlasse des Dr. A. Huber herausgegeben von Carl Brockelmann, Leiden 1892.
Goldziher, I., "Bemerkungen zur arabischen Trauerpoesie" in *Wiener Zeitschrift für die Kunde des Morgenlandes*, XVI, 1902 (= I. Goldziher, *Gesammelte Schriften*, 6 vols., (Hildesheim 1967-'73).
Guillaume, A., *The Life of Muhammad, a translation and notes*, Pakistan Branch Oxford University Press, Lahore Karachi Dacca 1967.
Ibn ʿAbbād, al-Ṣāḥib Abu l-Qāsim Ismāʿīl, *Al-kashf ʿan masāwiʾ shiʿr al-Mutanabbī*, ed. al-Shaykh Muḥammad Ḥasan Âl Yāsīn, Baghdad 1965.
Ibn ʿAbd Rabbih al-Andalusī, Abū ʿUmar Aḥmad ibn Muḥammad, *Kitāb al-ʿIqd al-farīd*, ed. Aḥmad Amīn, Aḥmad al-Zayn and Ibrāhīm al-Abyārī, 2nd printing, Cairo 1952-'53.
Ibn Hishām, *Al-Sīra al-nabawiyya*, ed. Muṣṭafā al-Saqqāʾ e.a., 2 vols., 2nd printing, Cairo 1955/ 1375.
Ibn al-Jarrāḥ, Abū ʿAbd Allāh Muḥammad ibn Dāwud, *al-Waraqa*, ed. ʿAbd al-Wahhāb ʿAzzām and ʿAbd al-Sattār Aḥmad Farrāj, 2nd printing, Cairo 1953.
Ibn Qutayba, *Kitāb al-shiʿr wa l-shuʿarāʾ*, Leiden 1904.
Ibn Rashīq, *Al-ʿumda fī maḥāsin al-shiʿr wa-ādābih wa-naqdih*, ed. M.M. ʿAbd al-Ḥamīd, 4th printing, Beirut 1972.
al-Iṣbahānī, Abu l-Faraj ʿAlī ibn al-Ḥusayn, *Kitāb al-aghānī* , ed. reprint Dār al-Kutub, (Cairo no date).
Jones, Alan, *Early Arabic Poetry vol. one: Marāthī and Ṣuʿlūk poems*, Edition, Translation and Commentary, Ithaca Press, Oxford University 1992.
al-Khaṭīb, Bushrā Muḥammad ʿAlī, *Al-rithāʾ fī shiʿr al-jāhilī wa-ṣadr al-islām*, Baghdad 1977.
Latham, J. Derek, "The Elegy on the Death of Abu Shujāʿ Fātik by al-Mutanabbī" in *Arabicus Felix: Luminosus Britannicus, Essays in Honour of A.F.L. Beeston on his Eightieth Birthday*, ed. Alan Jones, Ithaca Press, Oxford University 1991.
Lyall, Charles James, *Translations of ancient arabian poetry chiefly pre-islamic*, with an introduction and notes, London 1930.

Lyall, Charles James, *The Mufaḍḍalīyāt an anthology of ancient arabian odes, compiled by al-Mufaḍḍal son of Muḥammad according to the recension and with the commentary of al-Anbārī*, ed. Ch. J. Lyall, vol. II, Translation and notes, Oxford 1918.

al-Marzuqī, Abū ʿAlī Aḥmad ibn Muḥammad ibn al-Ḥasan, *Sharḥ Dīwān al-ḥamāsah*, ed. Aḥmad Amīn en ʿAbd al-Salām Hārun, vol. I-III, 2nd printing, Cairo 1967-'68, vol. IV, first print, Cairo 1953 /1373.

Nöldeke, Theodor, *Beiträge zur Kenntnis der Poesie der alten Araber*, Hannover 1864, reprint Hildesheim 1967.

al-Qālī al-Baghdādī, Abu ʿAlī, *Kitāb al-amālī*, 2 vols, ed. Dār al-kutub, Cairo, reprint Beirut, no date.

al-Rāghib al-Iṣbahānī, Abu l-Qāsim Ḥusayn ibn Muḥammad, *Muḥāḍarāt al-udabāʾ wa-muḥāwarāt al-shuʿarāʾ*, Beirut no date; in particular vol. 3 : 483-503 and 503-535, the chapter 'Mimmā jāʾa fī l-ghumūm wa l-ṣabr wa l-taʿāzī wa l-marāthī".

Rhodokanakis, N., "Al-Hansâʾ und ihre Trauerlieder, Ein literar-historischer Essay mit textkritischen Exkursen", in *Sitzungsberichte der Philosophisch-historischen Klasse der Kaiserlichen Akademie der Wissenschaften* 147ster Band (Vienna 1904), Abhandlung 4.

Schippers, A., "Abū Tammām's 'unofficial' elegies" in *Union Européenne des Arabisants et Islamisants, 10th Congress Edinburgh 9-16 September 1980*, Proceedings, ed. Robert Hillenbrand, Edinburgh 1982.

Shurūḥ Saqṭ al-zand li Abi l-ʿAlāʾ al-Maʿarrī, 5 vols., ed. Ṭaha Ḥusayn e.a., Cairo 1945-'48.

al-Tibrīzī, al-Khaṭīb, *Sharḥ ikhtiyār al-Mufaḍḍal, ṣanʿat al-Khaṭīb al-Tibrīzī*, ed. Fakhr al-Dīn Qabāwah, Damascus 1971-'72.

Wagner, E., *Grundzüge der klassischen arabischen Dichtung*, 2 vols., Darmstadt 1987.

Wellhausen, J., *Skizzen und Vorarbeiten*, Erstes Heft: 1. Abriss der Geschichte Israels und Juda's. 2. Lieder der Hudhailiten, Arabisch und Deutsch. Berlin 1884.

Winter, M., "Content and form in the elegies of al-Mutanabbī" in *Studia Orientalia D.H. Baneth Dedicata*, Jerusalem 1979.

Yalaoui, Mohammed, *Un poète chiite d'Occident au IVème / Xème siècle: 'Ibn Hânî' al-'Andalusî*, publications de l' Université de Tunis 1976.

Hebrew Andalusian Elegies and the Arabic Literary Tradition

Arie Schippers

INTRODUCTION

For a long time, Hebrew literature was essentially a part of a religious tradition. This, however, changed in the tenth century CE, when in Muslim Spain (al-Andalus) Jewish poets began to compose secular Hebrew poetry and inaugurated the 'Golden Age of Hebrew Andalusian[1] poetry' which reached its apogee under the Party kings (*Muluk al-tawa'if*) in the eleventh century[2]. The existence of several courts resulted in a competition that stimulated cultural life[3]. This emancipation of Hebrew literature can be explained by the special position the Jews occupied in al-Andalus, as compared with the other regions of the diaspora.

This special position was the result of several historical factors. In the first place, the relationship between the Muslim rulers and the Jewish community was a good one. The Jews, who had been living on the Iberian peninsula from the first century, had welcomed and even helped the Muslim conquerors in the eighth century. They saw them as their liberators since they had been oppressed by the Visigoth rulers. A second peculiarity was that Jews were not confined to certain professions but were to be found in all walks of life: among them were wage labourers, artisans, merchants and landowners. There were even exclusively Jewish cities such as Lucena and Granada. But perhaps most important for the development of Hebrew literature was the fact that there were Jews serving in high offices at Muslim courts. They often acted as Maecenases for Jewish scholarship and art.

---

1 I use the term Hebrew Andalusian poetry in contrast with Arabic Andalusian poetry. The term Andalusian Arabic poetry is occasionally used in contrast with Oriental Arabic poetry.

2 More information on the subject is to be found in Arie Schippers, *Arabic Tradition & Hebrew Innovation, Arabic Themes in Hebrew Andalusian Poetry* (Amsterdam, dissertation Institute for Modern Near Eastern Studies 1988), 290-338 and Arie Schippers, *Spanish Hebrew Poetry and the Arabic Literary Tradition*, (Leiden, E.J. Brill 1994), 244-286.

3 Muslim Spain was divided into several Muslim kingdoms led by ethnic Arabs, Berbers, Slavs, or Africans. In the eleventh century there was a rich cultural life at the courts, which competed with each other in wine drinking and poetry parties. Cf. for political history and the position of the Jews: Eli Ashtor, *The Jews of Moslem Spain*, (Philadelphia, Jewish Publication Society 1973, 1979, 1984), I-III; David Wasserstein, *The Rise and Fall of the Party-Kings; Politics and Society in Islamic Spain, 1002-1086*, (Princeton, University Press 1985); Raymond P. Scheindlin, "The Jews in Muslim Spain', in *The Legacy of Muslim Spain*, S.Kh. Jayyusi, ed., (Leiden, E.J.Brill 1992), 188-200.

One of the branches of Jewish scholarship that flourished in al-Andalus was Hebrew grammar. Hebrew had already disappeared in the second century as a spoken language, to be replaced first by Romance and later by vernacular Arabic. Inspired by the methods of the Arab grammarians who studied the Classical Arabic language, the focus of Hebrew grammatical studies was on the Classical Hebrew of the Holy Writ. These studies were greatly encouraged by the famous Maecenas and Cordoban vizier Hasday ibn Shaprut (ca. 910-970)[4].
The Hebrew poets tried to demonstrate that Classical Hebrew had the same possibilities as Classical Arabic for composing poetry and used it for correspondence in courtly circles and among friends, and for panegyrics. One of the first Hebrew poets to make a living out of poetry was Ibn Khalfun (ca. 970-ca. 1020)[5], who travelled around and sang the praise of high and influential Jewish statesmen and merchants. In this Hebrew Andalusian poetry the metres and themes[6] were adopted from Arabic poetry. Like in Arabic literature, we find here poems dedicated to wine, love, nature and war, and laudatory and elegiac poetry.

In the following, we will try to show the dependence of the Hebrew Andalusian poets on their Arabic examples, mostly Eastern Arabic poets. In doing so we will use examples from the four main poets of the Golden Age, namely Samuel ha-Nagid (993-1055)[7], Solomon ibn Gabirol (1021-1058)[8], Moses ibn Ezra (1055-1138)[9], and Yehudah ha-Lewi (1075-1141)[10]. We will show how these poets in the expression of their feelings about death and immortality made use of motifs and images borrowed from ancient Arabic poetic tradition. This does not imply that their feelings were not intense or serious enough. They just considered these Arabic themes and motifs, put into Classical Hebrew language, the best way to convey their feelings.

The main genre in Arabic literature is the *qasida*, a monorhymic and monometric poem of 20 to 60 lines. In Muslim Spain, Arabic poets also invented a strophic

---

4 See Angel Sáenz-Badillos and Judit Targarona Borrás, *Diccionario de autores judíos (Sefarad. Siglos X-XV),* (Córdoba, El Almendro 1988), 50-51.

5 See Sáenz Badillos, *Diccionario*, 162-163.

6 I use the term "themes' to indicate clusters of motifs, referring to contents and meaning; "genre" is used especially with regard to poetic forms. "Motifs' are small meaningful units.

7 Sáenz Badillo, *Diccionario*, 108-109; quotations from the poetry of Samuel han-Nagid according to the numbers of the poems: see Samuel han-Nagid [Shemuel Ibn Naghrila], *Diwan (Ben Tehillim)*, Ed. Dov Yarden, (Jerusalem 1966).

8 Sáenz Badillos, *Diccionario,* 91-93; quotations from Solomon Ibn Gabirol according to the numbers of the poems: see Solomon [Shelomoh] Ibn Gabirol, *Shire ha-Hol*, ed. H.Brody and J.Schirmann, (Jerusalem, 5735/1974) and id., *Shire ha-Hol*, ed. Dov Yarden, (Jerusalem 5735/1975).

9 Sáenz Badillos, *Diccionario,* 69-70; quotations from Moses [Moshe] Ibn Ezra according to the numbers of the poems: see Ibn Ezra , Moses [Moshe ibn Ezra], *Shire ha-Hol*, ed. Hayyim Brody, I, (Berlin 5695/1935); III (=comm.), Ed. D.Pagis, (Jerusalem 5738/1978).

10 Sáenz Badillos, *Diccionario,* 137-138; quotations from hal-Lewi according to the numbers of the poems: see Yehudah hal-Lewi, *Diwan*, Ed. H.Brody, (Berlin 5664/1904), II.

genre, consisting of several rhymes and metres, called *muwashshah* ('girdle poem', the girdle being the refrainlike end of each of five strophes)[11]. Both forms are also used in elegies. The themes and motifs of these elegies can be traced back to the sixth century poetry of the Arabian peninsula. Unlike the other poems, the elegiac poems did not begin with an amatory introduction (*nasib*). Instead, they began with the following motifs:

1. Descriptions of weeping and crying: hot tears and eyes tired because of sleepless nights. Sometimes 'reproachers' (*'awadhil*) intervene: now you have wept enough. Descriptions of sorrow and affliction. Participation of the universe (cosmos, animals, doves, other human beings) in the poet's grief[12].
2. 'Consolation' motifs: proverbs concerning the irreversibility of Fate[13], the transitoriness of earthly life, and the perfidy and faithlessness of the World. Every living being is doomed to die.
3. The weeping over the effaced traces of the grave of the deceased. The poet tries to make contact with the person who is buried, but to no avail. He addresses him, but gets no answer. This motif resembles the amatory introduction to other motifs of the *qasida*, where the poet wept at the abandoned campsite of his beloved.

In the middle of the elegies we find the following motifs:

4. Laudatory passages on the deceased.
5. Condolences by members of the family of the deceased; laudatory passages on the Maecenas.

At the end we often find:

6. Invocation of and address to the deceased person and benediction of him and the grave: 'May a rain fall upon it!' which is partly identical with the motif mentioned under 3.

---

11 Cf. Arie Schippers, "Style and Register in Arabic, Hebrew and Romance Strophic Poetry', in Federico Corriente and Angel Sáenz Badillos, eds., *Poesía estrófica, Actas del Primer Congreso Internacional sobre Poesía estrófica árabe y hebrea y sus paralelos romances (Madrid, diciembre de 1989)*, Federico Corriente and Angel Sáenz Badillos, eds. (Madrid 1991), 311-324.

12 See Ignaz Goldziher, "Bemerkungen zur arabischen Trauerpoesie', *Wiener Zeitschrift zur Kunde des Morgenlandes* (1902) = *Gesammelte Geschriften*, ed. Joseph Desomogyi (Hildesheim 1967-73), IV, 361-93; N. Rhodokanakis, *Al-Khansa' und ihre Trauerlieder,* (Vienna 1904); Ewald Wagner, *Grundzüge der klassischen arabischen Dichtung, Bd I, Die altarabische Dichtung, Darmstadt 1987; Bd. II Die arabische Dichtung in islamischer Zeit,* (Darmstadt 1987/88), I, 116-134; Arie Schippers, "Abu Tammam's Elegies on Ḫalid ibn Yazid al-Šaybāni', in *Festschrift Ewald Wagner*, ed. Gregor Schoeler, (Beirut 1994), II, 297-317. See also Pieter Smoor's article on this subject in the present collection; on the women's elegy, see Suzanne Pinckney Stetkevych, *The Mute Immortals Speak, Pre-Islamic Poetry and the Poetics of Ritual*, (Ithaca N.Y. & London, Cornell University Press, 1993), 161-205; and Gerard J. A. Borg, *Mit Poesie vertreibe ich den Kummer meines Herzens, Eine Studie zur Altarabischen Trauerklage der Frau,* (Nijmegen, dissertation 1994).

13 See Werner Caskel, *Das Schicksal in der altarabischen Poesie* (Leipzig 1926).

As far as formal features of the elegies are concerned we find through the whole poem:

7. Parallel structures, like internal rhyme, repetition of the name of the deceased, etc.

In order to show how the Hebrew Andalusian elegiac poetry is indebted to Arabic elegiac poetry[14] I shall deal in the following with four motifs: the participation of the whole universe in the grief, the consolation motifs, benediction formulas, and finally the impact of the deceased's status on the elegy.

PARTICIPATION OF THE UNIVERSE IN THE GRIEF

In the elaboration of this Arabic elegiac motif one finds a cosmic 'animism', which personifies as weeping entities sun and moon, stars, clouds, mountains, animals and the collectiveness of the world.
Samuel han-Nagid, in his (Hebrew Andalusian) elegy (poem no. 84) for the son of Rabbi Nissim, describes the participation of the universe in weeping in lines 5-6. It is as if common human beings do not know that heaven and its clouds are in distress because of his death, as are the crops in the field:

5. It is as if you [ungrateful human beings] did not know that heaven and its clouds are in great grief because of his calamity;
6. The harvest and vintage are in great pain, and the rain and the lightning experience disaster.

In poem no. 85 Samuel han-Nagid describes how the world is in confusion because of the death of Hay Ga'on (939-1038), the most important spiritual and juridical Jewish authority in Baghdad :

73. Alas! The sons of Earth are burning, and the World is in great confusion!
74. The inner moods are in the hands of the people; they are not doing their business.

When the stars hear about the execution, April 1039, of Abu Ishaq al-Mutawakkil ibn Hasan ibn Caprón, known as Yequtiel and famous as Shelomo ibn Gabirol's Maecenas in Saragossa and as a vizier of the Tujibid dynasty, they become extinguished, according to Solomon in his elegy (poem no. 156/194: line 27):

27. In darkness and night the stars have become dark while the days of Earth are cursed.

---

14 See on this subject Yehudah Ratzaby, "Arabic Influence on Hebrew Andalusian Elegies' [in Hebrew], in *Jerusalem Studies in Hebrew Literature* 10-11(1987-88)= *Essays in Memory of Dan Pagis*, 2: 737-765; Raymond P. Scheindlin, *Wine, Women and Death in Medieval Hebrew Poems on the Good Life* (Philadelphia, Jewish Publication Society 1986).

In an elegy for his brother Abu-l-Hajjaj Yosef (poem no. 117), Moses ibn Ezra describes the participation of the world in his weeping: the foundations of the world tremble because of the sad Fate of the poet; the stars of Heaven weep for him; moon and sun are grieving[15]. Similar descriptions can be found in the poems of Ibn 'Ammar and Ibn Zaydun[16]. He describes also how his heart trembles and his ears are deafened:

4. For me are mourning the stars of Heaven; and for me are weeping the moon and the sun;
5. Until daybreak my heart trembles, and my ear is uncovered and deafened by the noise of its voice.

In a poem (no. 229) on the death of Abu Ishaq Abraham, one of the sons of his sister, the poet describes his own grief and that of the universe:

28. The stars of Orion remove their splendour because of their death, while the sky's garment is a coat of darkness;
29. And the morning star is wrapped in darkness and howling, and every friend weeps for his friend;
30. If there could be made a ransom for a living being, then the Great Bear would make a ransom of its own children.

'CONSOLATION' MOTIFS: THE IRREVERSIBILITY OF FATE, THE PERFIDY OF THE WORLD.THE TRANSITORINESS OF LIFE ('EVERY LIVING BEING IS DOOMED TO DIE')

Before Islam the Arab pagan poets saw blind, irreversible Fate as the dominating power in their lives. This notion figured prominently in Arabic elegiac poetry. Life in this World was called treacherous, since no one knew what it had in store for him. Since the coming of Islam, these pre-Islamic notions continued to have their place in elegies, sometimes juxtaposed with Islamic concepts. This could result in a certain scepticism, as expressed in an elegy composed by the Arabic poet al-Mutanabbi[17]:

43. People are in disagreement about death, and there is only consensus about the fact of death.

---

15 Mercedes Etreros and Angeles Navarro, "Mošeh ben 'Ezra; elegías a la muerte de su hermano José", *Misceláneas de Estudios Arabes y Hebraicos* 32, 2 (1983), 51-68.

16 See Arie Schippers, "Two Andalusian poets on exile: Reflexions on the poetry of Ibn 'Ammar (1031-1086) and Moses ibn Ezra (1055-1138)", in *The Challenge of the Middle East*, I.A. El-Sheikh, C.A. van de Koppel, Rudolph Peters,eds. (Amsterdam 1982), 113-121, especially 114.

17 Elegy on the elder sister of Sayf al-Dawla, see Abu l-Tayyib al-Mutanabbi, *Diwan*, ed. Fridericus Dieterici, (Berlin 1861), poem no. 238.

44. Some say man's soul survives safe and sound, and others say it shares the destruction of the body.[18]

Hebrew Andalusian poets also made use of this motif of blind Fate, to which they usually referred with expressions such as 'Time', or 'the Days' and 'the Nights'. In an elegy on a friend (poem no. 105), Samuel han-Nagid, comforting the deceased's relatives, discusses the untrustworthiness and unreliability of Life and Earth: benefactors die as much as fools. Life is a bad dream. Life and Death are two quarreling women. People give birth to sons; Time gives birth to catastrophes, which annihilate human beings. There is hardly a son that does not witness his father's death, but the father is also deprived of his sons (lines 1-9). Part of the text runs as follows:

5. Life and Death, there is no honest justice between them: they judge every soul;
6. Between them is hostility, like the hostility of two quarreling women, who have nevertheless one husband;
7. This man destroys what his friend has built, like wrongdoers, but they are no young boys (urchins);
8. Mankind gives birth and Time generates: the sons of man are to be eaten by the sons of Time;
9. If the son does not behold the death of [his] father: at least then the sons of the father, during the life of their father, are bereft [of their children].

Similar thoughts are expressed by Moses ibn Ezra (in poem no. 117): innocents and sinners have the same fate (lines 45-53):

48. The innocent and the impious are given to drink from one and the same cup, and their fate is like the fate of beasts;
49. A man is born naked and will go away when his soul is destitute of the World's labour.
50. He will leave his functions behind him for another and cannot take anything, even the slightest, with him on the day that he shall go.
51. His deeds accompany him to his grave to give him rest, when his behaviour has been perfect.
52. The car in which the man goes to his grave to annihilation will be harnessed from [his birth];
53. Death is indeed like a sickle, and we are like the harvest which is ripe and standing.

---

18 Translation M. Winter, "Content and Form in the Elegies of al-Mutanabbi," in *Studia Orientalia Memoriae D.H. Baneth Dedicata*, (Jerusalem 5739/1979), 327-345 especially 342.

In another poem (no. 229)[19] Moses ibn Ezra speaks about unjust Time, and he mentions the general character of Death. The World is transitory and untrustworthy (lines 44-54), although the young Abraham's only interest is in holy and durable things (lines 55-57). The following lines forcefully express the transitoriness of life:

51. The cup of Death is mixed in the hands of the Days, and drinking of its wine is a custom for every living being.
53. O man! Of what avail to you is the pursuit of honour in the World! What use to you are the riches you have gathered?
54. What benefit is there for you to extract honey from bitterness: after a while you will vomit it up.

The 'Life is a Dream' motif that we find occasionally in Arabic Andalusian poetry occurs frequently in Hebrew Andalusian elegies. Samuel ha-Nagid says e.g. in his poem no. 30: line 13[20]:

13. Your World is like a dream, and in your old age you find many of its explanations; in your Death you will find all its solutions.

In Solomon ibn Gabirol's elegy (poem no.156/194) on his above-mentioned Maecenas Yequti'el, we find the following consolation motif combined with a criticism of all people who believe in slogans like 'Life is a Dream': the fact that Yequti'el's days now have come to an end indicates that the cosmos has been created in order to perish. The poet adjures mankind to pay attention to the fact that Time had already, before their birth, arranged their grave. The Days of Time give here and take there, so that you are led to believe that men are merchandise (lines 1-3):

2. Pay attention in order that you may know that Time arranges graves for people, when they are not even born!
3. Its Days take people and they give other people, so that I think that they make bargains with mankind.
4. I do not know it, just as all the wise people whose names are known did not know ...
5. Whether the souls despised the bodies or the living spirits are urged to go up.
6. How foolish are those who say: 'The World is a dream!' How could they have forgotten a thing and not remember it any more.

---

19 An elegy on the death of his nephew (son of his sister): *wa-nuʿiya ilay-hi Abu Ishaq ibn ukhti-hi tuwuffiya shenat* [the death of Abu Ishaq, his sister's son, was announced to him in the year] 4882 = 1122.

20 See also Samuel han-Nagid, *Ben Qohelet*, ed. S. Abramson, (Tel Aviv 5713/1953), no. 272, which is identical.

Like al-Mutanabbi in his elegy, quoted before, Ibn Gabirol expresses the uncertainties of the mysteries of Death, which human beings cannot solve. Dreams are vain and cannot be explained (line 7)[21].

The poet continues with his consolation concerning the World: the World has arranged an ambush for the fatigue [of mankind] (line 8). It is interesting that the poet speaks about the covenant of the kings of the earth with Death or the Underworld: but this covenant has now become void and without value[22]. Prophets summoned mankind to show repentance, but the people did not respond. Princes as well as servants have to descend suddenly to the grave, without regard for their state. Princes cannot take their riches: they have to leave behind them everything except the rags in which they are buried. They are locked up under the earth (lines 9-17). The poet addresses himself to Death. He looks just, because he redeems the poor from their life and makes rich people stumble. Even if the mighty ones mock at you, they become like nails from terror. If the fire falls upon the cedars, how can the mosses on the wall escape from it? (lines 18-21). The poet does not know why the Days have despised him and why they have chosen precisely him. They have taken away the chosen people of the earth. The poet knows that the deeds of the Days have given birth to calamities. He knows that They have sinned consciously. Do not trust the Days, do not make a treaty with Them, although they look like the Angel of the Covenant. If the Days have promised to annihilate all precious things, they have performed their promise (lines 28-32).

World and Time are described basically as treacherous in Hebrew Andalusian poetry. Moses ibn Ezra seems to think in an elegy on his brother Yosef (poem no. 117) that Time apparently has wanted to take revenge on him (lines 7-14): Time wants to destroy all the sons of his father. It does not spare even the little boys and girls of his family, whereas, in former times, Time used to serve the poet humbly:

9. In former times It ran to comply with the wishes of my heart and my longings as a slave or a maid-servant;
10. Such is Its custom in making Itself odious to the innocent and making pacts with men of blood and treason.
11. Its favour is abundant for every rogue, but its [scant] measure for every generous person is abominable.
12. Every innocent walks barefoot, while every brother of evil washes his feet in oil;
13. Time has decided to let me live, but to let my brothers die: how mendaciously It has behaved, full of treason;
14. It has given me to eat the bitterness of its wrath and to drink the venom of its anger and fury.

Here the poet stresses the injustice of Time, taking away the innocent, whereas the wicked people are allowed to stay alive. The poet considers also that his brothers deserved more to stay alive than he himself.

---

21 Winter, "Content', 342.

The motif of the treacherous World by Moses ibn Ezra can also be found in another poem (no. 141):

1. The one who puts to the test treacherous Time will meet in it a beloved one who hides the mantle of the hater he is in reality.

In Arabic poetry e.g. by al-Mutanabbi, we can observe that the poet imagines that his Maecenas had in vain tried to make a deal with Fate[23]. In Hebrew Andalusian literature we see this motif expressed by both Samuel han-Nagid and Moses ibn Ezra. In his *Muhadarah*[24] (129b) Moses ibn Ezra quotes as an example of *tasdir*[25] the following line:

> He thinks that Time will perform his will unto a good end, but [Time] itself does not think so.

In panegyrics Fate, Time, the Days and their vicissitudes stay in opposition to the Maecenas, the just and mighty ruler, to whom Fate is subdued. In elegies, Fate and its calamities have won. Fate here has to be considered in opposition to God, whose benediction is asked on the grave by means of a raincloud which drenches the grave grounds.

BENEDICTION FORMULAS

The early Islamic poet Mutammim ibn Nuwayra asks God to send rain clouds to the grave of his brother Malik[26]. The benediction is asked from God:

23. I say, while the flashes light the cloud-fringe below the black, and down from the dark mass pours the rain in a scattering flood:
24. 'May God grant the land where lies the grave of my brother dear the blessings of showers at dawn, and cover the ground with green.'

The same motif we find in Arabic Andalusian poetry:

---

22 Ibidem.

23 Schippers, *Arabic Tradition*, 311; Schippers, *Spanish-Hebrew Poetry*, 262: 'Fate has made a deal with Sayf al-Dawlah by taking away the younger sister and sparing the elder one [poem no. 231 of al-Mutanabbi's *Diwan* (ed. Dieterici), lines 12-13] [..]. But Time deceives Sayf al-Dawlah, as we see in the elegy on the elder sister Khawlah (no. 238) because now Death has also taken Khawlah.' See also Winter, "Content", 342.

24 Moses Ibn Ezra [Moshe ibn Ezra], *Kitab al-Muhadara wa-l-Mudhakara*, ed. A.S. Halkin, (Jerusalem 5735/1975); and ed. Montserrat Abumalham Mas, (Madrid, Consejo Superior de Investigaciones Científicas, 1985 [I: edición], 1986 [II: traducción], 129b (quoted according to Bodleiana Neubauer Ms. no. 1974, ff. ab).

25 Beginning and ending the line with the same word.

26 Al-Mufaddal al-Dabbi, *The Mufaddaliyyat: An Anthology of Ancient Arabian Odes, ed. Charles James Lyall,* I+II (translation and notes), (Oxford 1918), poem no. 67: line 24 .

1. 'O my brother Husayn, unique being of Time, may God water your grave with the shower of a spring rain.'[27]
23. May God water a rebellious grave at Sfax with rich rainclouds which satisfy the earth.[28]
7. May God drench the grave of my father with mercy, so that a fragrant morning cloud will be his shower.[29]

In Hebrew Andalusian poetry this motif is used too as the following examples show:

26. May He drench his grave with clouds, may He shed morning showers on his dust.[30]
3. May He (the Rock= God) drench the rock of her grave with the waters of a cloud so that she may never fear a desert. [31]
50. May a cloud drip to drench her grave with the waters of good will, and may a layer of dew rise early in the morning.[32]
39. May her grave flower like an irrigated garden and be watered at every moment with the dew of the morning.[33]

From the examples quoted above, it will be clear that both in the Arabic and Hebrew Andalusian elegiac poetry, several pre-Islamic concepts have inspired the most commonly used motifs: cosmic animism, fatalism, and the belief in a high God, which in pre-Islamic times was not necessarily the only one. These three different religious elements are juxtaposed. The use of the pre-Islamic concepts gives the Hebrew Andalusian elegiac poetry an archaic flavour.

THE STATUS OF THE DECEASED

Naturally, the status of the deceased greatly affected the contents and style of the elegy. As in Arabic poetry, important leaders and learned men, if they are the object of an elegy, are addressed in a great and magnificent manner. But if the deceased was a relative or friend, the poem as a rule is shorter and more informal. Women and children offered some problems to the poet, since it was difficult to deal with their personal lives: this was true for women, since they did not participate in public life, and for children because they had not left the protection of the family circle and achieved memorable feats.

---

27 Muhammad Ibn 'Abd al-Karim (8th/ 14th century), see Lisan al-Din Ibn al-Khatib, *Al-Katiba al-Kamina,* ed. Ihsan 'Abbas (Beyrouth 1979), 213.
28 Ibn Hamdis, *Diwan*, ed. Ihsan "Abbas, (Beyrouth 1960/1379), 35 (poem no. 28).
29 Ibn Hamdis, *Diwan* , 522 (poem no. 330).
30 Samuel han-Nagid, *Diwan* , poem no. 105.
31 Moses ibn Ezra, *Diwan* , poem no. 3 on Abu Yahya Ibn al-Rabb's wife.
32 Moses ibn Ezra, *Diwan* , poem no. 137 on the mother of the Banu Mashkaran.
33 Moses ibn Ezra, *Diwan* , poem no. 53 on Yoshiah ibn Bazzaz's sister.

In an elegy for a young child of Rabbi Nissim, Samuel han-Nagid comforts the father[34]. We see in this poem some of the same motifs that are present in the elegy for Abu-l-Hayja' by al-Mutanabbi[35]. In a passage in which he praises the father rabbi Nissim[36], he says: the father has met with a paradoxical disaster. His only [*yahid*] child has been nipped in the bud, while the father is a unique [*yahid*] person in the world. In the time after the period of lactation, his sprig has been uprooted, although its root was a king (i.e. rabbi Nissim), who granted abundant gifts; he who supported those who were oppressed and stumbled, has now stumbled himself because of the loss of his offspring. He has not even had the pleasure of teaching him with a slate [*luah*], he who used to bear with him the Commandments [*luhot*] of the Covenant as bracelets. If, of all children, he, rather than another child, has had to die, what kind of pride can fathers have in their children any more? (lines 7-11).

Most of Samuel han-Nagid's elegies are devoted to the death of his brother Yishaq Abu Ibrahim. His elegies on family members give us a good idea of how people of his time dealt with distress and mourning. Yishaq's death moved him deeply, in view of the many poems which he dedicated to him: nearly every stage, from death-bed to burial, is described in a separate poem. The first elegy (poem no. 86) was composed when Yishaq was lying on his deathbed (in the year 441/ 1050/ 4810). The elegies for Yishaq are mostly very short. The disease worsens, and in the next poem the poet tells of his desire to pay a visit to his brother accompanied by the physician, but halfway they meet a man who announces that Yishaq has just died (no. 87). Other poems follow. One of these deals with the tearing of his clothes as a token of his mourning (no. 89).

Then comes a long poem in which he calls to memory his brother's death, and the passing away of three other family members (no. 90). A short poem relates his entry into the funeral parlour, where his brother is lying in state (no. 91). The poet describes kissing his brother. Poem no. 92 describes how his brother is dressed for the grave and how he is put into the grave. In poem no. 93 the poet comforts people who are weeping and mourning because of his brother's death; then follow two poems composed after a visit to his brother's grave on the second day after the burial (nos. 94, 95). Then follows a poem after his return from the grave, when he arrives at his working place, surrounded by a multitude of people (no. 96).

---

34 Moses ibn Ezra, *Diwan* , poem no. 84.

35 Arie Schippers, "Abu Tammam's unofficial elegies," *Acts of the 10th Congress U.E.A.I. Edinburgh 1980* (Edinburgh 1982), 101-106, especially 104: 'It is amusing to see how al-Mutanabbi succeeds in describing the virtues of this child, who did not even reach the age when he could walk, but for whom enemies were already trembling in view of his future bravery.'

36 For Nissim ibn Shahin who lived in Tunesia, see *Encyclopedia Judaica* (Jerusalem, Kether 1970),12: 1183.

Poems nos. 97, 98, 100 and 103 remember Yishaq's death, respectively, a week, a month, and a year after his death. There is also a poem about his going to the grave, which is interrupted half-way (no. 99). Samuel han-Nagid describes in poems nos. 94 and 99 how he unsuccessfully tries to make contact with his deceased brother:

94:1. Alas! I returned home in the distress of my soul, may God show favour to you, O my brother!
2. I buried you in the grave yesterday and even to-day is my complaint bitter (Job 23:2);
3. Peace be with you! Don't you hear that I am calling you with all my forces?
4. Answer me: do you recognize the answer to my elegies by my crying?
5. How are you lodging in your grave on the dust in the house of ruins?
6. Are your bones already becoming weak, and are your teeth in your face removed from their place?
7. Has your freshness flown away in one night, for my freshness has gone away because of the weeping;
8. I have left you, O elder son of my father, as a deposit in the hand of the One who has produced me;
9. I trust in the One who made my hope, that you will go unto peace (Psalm 22:10[9]).

99:1. Is there a sea between me and you, that I should not turn aside to be with you;
2. That I should not run with a troubled heart to sit at your grave-side?
3. Truly, if I did not not do so, I would be a traitor to our brotherly love!
4. O my brother, here I am, facing you, sitting by your grave;
5. And the grief for you in my heart is as great as on the day you died;
6. If I greeted you, I would hear no reply.
7. You do not come out to meet me when I visit your grounds.
8. You will not laugh in my company, nor I in yours.
9. You cannot see my face, nor I yours,
10. For the pit is your home, the grave your dwelling-place!
11. First-born of my father, son of my mother, may you have peace in your final rest,
12. And may the spirit of God rest upon your spirit and your soul!
13. I am returning to my own soil, for you have been locked under the soil.
14. Sometimes I shall sleep, sometimes wake – while you lie in your sleep forever.
15. But until my last day, the fire of your loss will remain in my heart!

Samuel han-Nagid also composed an elegy about the books which will not appear any more since his brother Yishaq is dead (no. 101). Whenever the poet sees other people mourning a family member, he is prompted to weep again over the loss of his brother (no. 102). He also composed a consolation poem devoted to his brother's death (no.104). Of all his elegies and mourning poems (24 to 26) 19 deal with his brother.

Moses ibn Ezra composed elegies for his brother Abu-l-Hajjaj Yusuf or Yosef. In one of these elegies (no. 113), he manifests his great affliction concerning his brother, whose glory will remain forever, although his body is decomposing (lines 26-27). In this elegy the poet describes how far away he is from his brother; he addresses his brother in the following way (no. 113):

28. Alas! I wish I could save you, O my brother, but Death will strike me before I will come near to you;
29. I was in my lifetime in exile because I was separated from you; now I want a grave as a part of your inheritance;
30. I call to [the grave] to a distant person: ' I wish I had a grave 'and then I dig my grave next to you.

In some of the short mourning pieces of Moses ibn Ezra, we can see the embarrassment of a poet when the deceased is a young boy, an embarrassment similar to that to be found in Abu Tammam's poem for the young boy Humayd[37]. The death of Moses ibn Ezra's young son Yaʿqob inspired four small poems (nos. 32, 97, 135, 171). These poems are simple in tone, as can be seen from the following examples; no. 32 runs as follows:

1. The name of Yaʿqob has perished, but the plant of his calamity has increased in my heart;
2. I weep in order to annihilate it, but it increases instead! Because a plant grows immensely by watering.

In the poem little is said about Yaʿqob: it consists only in describing the father's sadness by means of a rhetorical device, called in Arabic *husn al-ta ʿlil* or 'good argumentation', which indicates a kind of fantastic paradox. Other elegies on his little son go as follows:

97:1. My heart went away to behold Yaʿqob: he travelled dead on the arms of his bearers;
2. Apart from them we shall never see anyone in the world who buried a star in dust.

135:1. The day on which Yaʿqob expired, the sound of my violin turned to mourning;

37 Schippers "Abu Tammam's unofficial elegies", 103: '1. Humayd is dead. But what soul remains on earth without dying?; 2. I weep for him with tears in my eye like scattered pearls. 3. He was but an unexperienced boy, when Fate made him ill. I will never forget him, as long as I live. 4. It is impossible to find any praise for his qualities, so the best thing I can do is maintain silence.'

2. As for that day, let darkness seize upon it; may this day of evil and bitterness be annihilated !

171:1. My tears flow from the warmth of my heart, whose twelve rags are torn;
2. It is a constant duty to weep for a dear son, a child who was gladness.

These three poems have almost no laudatory motif, but only mourning and weeping, sometimes with clear reminiscences of expressions of Biblical language, e.g. poem 135 which reminds us of Job 30:33 and 3:6 and – to a lesser extent – Amos 8:10 and Jeremiah 2:19.
As for poem 97, here there is some laudatory motif: the heavenly qualities of the deceased are contrasted with the dust in which he is buried.

One of the most delicate elegies on a woman by Yehudah hal-Lewi is the one on his little daughter: in elegy no. 33 (lines 16-23), the poet describes the deceased girl as a rose or a lily plucked before her time, whose image is before the eyes of the poet like phylacteries. How his tears flow like a river! How can it be that the daughter of the constellation of the Great Bear is now gathered in the womb of the dust! How can a sun revolve with the worms, while the cords of the subterranean vault are a diadem for her head? Then this verse is closed with the usual refrain in which the poet repeats: 'O daughter, there is no judge who can help in any way with your case, because Death separates you from me'. In this poem – as in other elegies we have seen – the contrast between her heavenly qualities and her actual subterranean state is made again.
The same poem contains a dialogue between the mother and the daughter, described by the poet[38]:

1. 'Alas, my daughter, have you forgotten your home?
2. The coffin bearers have taken you to the grave,
3. And I have nothing left of you but your memory.
4. I take pity on the dust of your tomb,
5. When I come to greet you, and do not find you:
6. For Death has made a separation between me and you'.[...]
33. 'Alas, my daughter, what sorrow you have brought me!'
34. 'Alas, alas! my mother, that you ever gave me birth;
35. How, on this day, how could you cast me off?
36. O, you brought me up to be Death's bride!
37. When my turn came, you sent me away alone;
38. You crowned me with a garland of dust;
39. You set me down in the bridal-bower of destruction.
40. It was against your will, O my mother, it was not according to your will;
41. Because Death has made a separation between me and you.'

---

38 See T. Carmi, *The Penguin Book of Hebrew Verse*, (Harmondsworth 1981), 339-40.

Elegies on a daughter were not uncommon in the Arabic world. In a recent article and lecture, T. Emil Homerin[39] gives us a whole series of elegies on a daughter of an Egyptian religious scholar Abu Hayyan, who also tried to contact the girl once she was buried in the grave. In his article, Homerin also refers to some elegies on daughters in previous periods in Classical Arabic literature[40].

The quotations in this passage are all from 'personal' elegies which show more intimacy than the more solemn and longer elegies on great viziers and scholars such as Hay Ga'on and Yequti'el, from which we have quoted some samples in previous sections. However, in one respect there is a difference with Arabic elegies on important men: Jewish viziers are almost exclusively praised because of their intellectual capacities, and not because of their bravery in warfare, as in the case of their Arab counterparts. Both, however, are praised for their generosity. There is yet another difference. Hebrew Andalusian poetry is more often addressed to equals than Arabic poetry. This is due to the fact that the latter was often written for persons of higher standing in the expectation of remuneration. Hebrew poetry, on the contrary, was often used in correspondence between rabbis, that is persons of equal status, who often were friends. Generosity in Hebrew poetry then does not always refer to material rewards such as money, precious gifts or beautiful slave girls, but could also refer to immaterial goods, such as a poem.

CONCLUSIONS

When dealing with the elegiac genre in world literature, one is struck by the universality of this genre. In other literatures, for instance, Medieval Provençal poetry, one can also find thematic sequences such as 1) invitation to lamentation; 2) praising the deceased one and his lineage; 3) enumeration of countries and persons distressed by his death; 4) praising his virtues; 5) benediction, greeting, or prayer for salvation; and 6) description of distress[41]. This superficial likeness of elegiac thematics is not accidental. Behind the poetic experience lies the actual event which was the incentive towards the composition of the poem. In a still

---

39 Th. E. Homerin, "A Bird Ascends the Night, Elegy and Immortality in Islam," *Journal of the American Academy of Religion* 58, 247-279.

40 To a quite strange category of elegies belongs the one which Yehudah hal-Lewi made for himself (no. 32): it does not belong to the category of praise, because the poet mainly describes how the angel of Death came suddenly to him, without even permitting him to make his own testament or to take leave of his own sons and daughters. Unlike the few Arabic examples of elegies in this genre, there is no self-praise in it. See Ibrahim al-Hawi, *Ritha' al-Nafs bayn ʿAbd Yaghuth ibn Waqqas wa-Malik ibn al-Rayb al-Tamimi*, (Beyrouth, Risala 1988); see for Abu Nuwas' elegies on himself, Ewald Wagner, *Abu Nuwas, eine Studie zur arabischen Literatur der frühen Abbasidenzeit* (Wiesbaden, 1965), 359.

41 Cf. C. Cohen, " Les éléments constitutifs de quelques planctus des Xe et XIe siècles," *Cahiers de civilisation médiévale*, I, 1958, 83-86.

unpublished lecture James Montgomery[42] looks for the ties between actual experience and poetic experience, and traces in the poetic sequences the states of mind which, according to psychologists, people in distress go through: such as denial of the event, anger, bargaining with Fate or God, depression and resignation, which are also easy to find as poetic topoi in the elegies. Nevertheless, however universal the genre may be, it is striking that in Arabic and Hebrew Andalusian elegies themes and motifs are connected with pre-Islamic concepts which have survived for centuries, such as the concept of Fate which we stressed earlier. How universal this elegiac genre may be, in the elaboration of its motifs, the poets still maintained its specific archaic pre-Islamic background. This does not mean that the Hebrew Andalusian elegies are poems without originality and ingenuity and that the Hebrew poets are mere copyists of the Arabic tradition. Every poem and every poet has, although within the tradition, its own individuality and character and own use of the Hebrew Classical language. Within this elegiac tradition he finds the best way to express his feelings about death and mortality. It is as if, before the mysteries of Death and Immortality, with so many insecurities and doubts, the poet wants to cling to something more sure and stable, to old expressions and language which reflect archaic manners and customs. Those expressions have proved themselves during centuries. We, also, people of the twentieth century, express ourselves only by stereotyped ideas, when suddenly confronted with someone's death.

---

42 James Montgomery, "Al-Mutanabbi's threnody for Abu Shuja al-Fatik and the psychology of grief," Lecture at the *1993 Cambridge Symposium of Classical Arabic Poetry* (forthcoming).

BIBLIOGRAPHY

Eli Ashtor, *The Jews of Moslem Spain*, (Philadelphia, Jewish Publication Society 1973, 1979, 1984) I-III.

Werner Caskel, *Das Schicksal in der altarabischen Poesie*, (Leipzig 1926)

Al-Mufaddal al-Dabbi, *The Mufaddaliyyat: An Anthology of Ancient Arabian Odes, ed. Charles James Lyall,* I+II (translation and notes), (Oxford 1918)

J. Derek Latham , "The Elegy on the Death of Abū Shujāʿ Fatik by al-Mutanabbi," in Alan Jones, ed., *Arabicus Felix, Luminosus Britannicus, Essays in Honour of A.F.L. Beeston on his Eightieth Birthday*, (Reading, Ithaca Press 1991), pp. 90-107.

Encyclopedia Judaica (Jerusalem, Kether 1970) I-XIII.

Mercedes Etreros and Angeles Navarro, "Mošeh ben 'Ezra; elegías a la muerte de su hermano José," *Misceláneas de Estudios Arabes y Hebraicos* 32, 2 (1983), 51-68.

Th. E. Homerin, "A Bird Ascends the Night, Elegy and Immortality in Islam," *Journal of the American Academy of Religion* 58, 247-279.

Moses [Moshe] Ibn Ezra, *Shire ha-Hol*, ed. Hayyim Brody, I (Berlin 5695/1935); III (=comm.), Ed. D.Pagis, (Jerusalem 5738/1978).

Moses Ibn Ezra [Moshe ibn Ezra], *Kitab al-Muhadara wa-l-Mudhakara*, ed. A.S. Halkin, (Jerusalem 5735/1975);

id., ed. Montserrat Abumalham Mas, (Madrid, Consejo Superior de Investigaciones Científicas, 1985 [I: edición], 1986 [II: traducción] (both quoted according to Bodleiana Neubauer Ms. no. 1974, ff. ab).

Solomon [Shelomoh] Ibn Gabirol, *Shire ha-Hol*, ed. H.Brody and J.Schirmann, (Jerusalem, 5735/1974)

id., *Shire ha-Hol*, ed. Dov Yarden, (Jerusalem 5735/1975).

Ignaz Goldziher, "Bemerkungen zur arabischen Trauerpoesie," *Wiener Zeitschrift zur Kunde des Morgenlandes* (1902) = *Gesammelte Geschriften*, ed. Joseph Desomogyi (Hildesheim 1967-73), IV, 361-93.

Husayn Yusuf Khuraywish, "Dirasat al-janib al-fanni fi-l-marthiya al-Andalusiyya," in *Maʿrifa* 18, 216 (1980), 99- 121.

Yehudah hal-Lewi, *Diwan*, Ed. H.Brody, (Berlin 5664/1904), II.

James Montgomery, "Al-Mutanabbi's threnody for abu Shuja al-Fatik and the psychology of grief," Lecture at the *1993 Cambridge Symposium of Classical Arabic Poetry* (forthcoming)

Abu -l-Tayyib al-Mutanabbi, *Diwan*, ed. Fridericus Dieterici, (Berlin 1861)

Samuel han-Nagid, *Ben Qohelet*, ed. S. Abramson, (Tel Aviv 5713/1953

Samuel [Shemuel]han-Nagid Ibn Naghrilah, *Diwan (Ben Tehillim)*, ed. Dov Yarden (Jerusalem 1966).

Yehudah Ratzaby, 'Arabic Influence on Hebrew Andalusian Elegies' [in Hebrew], in *Jerusalem Studies in Hebrew Literature* 10-11(1987-88)= *Essays in Memory of Dan Pagis*, 2: 737-765.

N. Rhodokanakis, *Al-Khansa' und ihre Trauerlieder,* (Vienna 1904).

Angel Sáenz-Badillos and Judit Targarona Borrás, *Diccionario de autores judíos (Sefarad. Siglos X-XV),* (Córdoba, El Almendro 1988).

Raymond P. Scheindlin, *Wine, Women and Death in Medieval Hebrew Poems on the Good Life*, (Philadelphia, Jewish Publication Society 1986).

Raymond P. Scheindlin, "The Jews in Muslim Spain," in *The Legacy of Muslim Spain*, S.Kh. Jayyusi, ed., (Leiden, E.J.Brill 1992), 188-200.

Arie Schippers, "Two Andalusian poets on exile: Reflexions on the poetry of Ibn 'Ammar (1031-1086) and Moses ibn Ezra (1055-1138)," in *The Challenge of the Middle East*, I.A. El-Sheikh, C.A. van de Koppel, Rudolph Peters,eds. (Amsterdam 1982), 113-121.

Arie Schippers, "Abu Tammam's unofficial elegies," *Acts of the 10th Congress U.E.A.I. Edinburgh 1980*, (Edinburgh 1982), 101-106,

Arie Schippers, "Style and Register in Arabic, Hebrew and Romance Strophic Poetry," in Federico Corriente and Angel Sáenz Badillos, eds., *Poesía estrófica, Actas del Primer Congreso Internacional sobre Poesía estrófica árabe y hebrea y sus paralelos romances,(Madrid, diciembre de 1989)*, Federico Corriente and Angel Sáenz Badillos, eds. (Madrid 1991), 311-324.

Arie Schippers, *Arabic Tradition & Hebrew Innovation, Arabic Themes in Hebrew Andalusian Poetry* (Amsterdam, dissertation Institute for Modern Near Eastern Studies 1988).

Arie Schippers, *Spanish Hebrew Poetry and the Arabic Literary Tradition*, (Leiden, E.J. Brill 1994).

Arie Schippers, "Abu Tammam's Elegies on Ḫalid ibn Yazid al-Šaybāni," in *Festschrift Ewald Wagner zum 65. Geburtstag, (=Beiruter Texte und Studien , Band 54)*, ed. Wolfhart Heinrichs & Gregor Schoeler, (Beirut, Stuttgart, Franz Steiner 1994), II [=Studien zur arabischen Dichtung], 297-317.

Pieter Smoor, *see his article in the present collection.*

Suzanne Pinckney Stetkevych, *The Mute Immortals Speak, Pre-Islamic Poetry and the Poetics of Ritual*, (Ithaca N.Y. & London, Cornell University Press, 1993), 161-205

Ewald Wagner, *Grundzüge der klassischen arabischen Dichtung, Bd I, Die altarabische Dichtung, Darmstadt 1987; Bd. II Die arabische Dichtung in islamischer Zeit,* (Darmstadt 1987/88), I, 116-134.

David Wasserstein, *The Rise and Fall of the Party-Kings; Politics and Society in Islamic Spain, 1002-1086*, (Princeton, University Press 1985).

M. Winter, "Content and Form in the Elegies of al-Mutanabbi," in *Studia Orientalia Memoriae D.H. Baneth Dedicata*, (Jerusalem 5739/1979), 327-345.

PHILOSOPHICAL

Death and Immortality in Greek Philosophy

From the Presocratics to the Hellenistic Era*

Bartel Poortman

When in his Nicomachean Ethics Aristotle describes death as φοβερώτατον[1], he seems to be expressing a view that was commonly held. Similar statements by Socrates[2] and Epicurus[3] show that, then as now, man, who, as Aristotle points out[4], *knows* he will die, dreaded death; either through fear of dissolution, the state of death as eternal non-existence, or through the gloomy prospect offered by the traditional view of survival of *psuchè*[5] in Hades, where punishments might be waiting and which, given Achilles'[6] well-known complaint, was a dismal place to stay anyway.

Although in Schopenhauer's view philosophy would barely be possible in the absence of death, indeed, death is "der eigentliche inspirierende Genius oder der Musaget der Philosophie",[7] it was not until Plato (427-347) that death and its implications effectively became the subject of philosophical discussion.

It may seem surprising that, unlike Greek literature and the visual arts,[8] in which death figured prominently from the beginning, philosophy took comparatively long before paying regard to something so inextricably entwined with life as death, but considering the nature of early Greek philosophy this is quite understandable. The Presocratic philosophers were primarily engaged in cosmological enquiries, in which man as such and problems immediately connected with the human condition were of marginal importance or did not come in at all. In the initial stage, attempts to escape the finitude of human existence by offering the prospect of personal immortality should, therefore, not be looked for in a philo-

---

* I should like to thank my colleague Marga Jager, who translated the greater part of this article into English and who made a number of valuable critical comments, and Mrs C. Crouwel-Bradshaw, who corrected my own translation of the remaining part.

1 *E.N.* III.vi, 1115a26.

2 Plato, *Apol.* 29a8-b1.

3 Diogenes Laertius X.125.

4 *Rhet.* II.iv, 1382a26-7.

5 In view of the religious connotations of the modern word 'soul' I prefer to transcribe ψυχή by *psuchè*.

6 Hom., *Odyss.* 11.488 sqq.

7 A. Schopenhauer, *Die Welt als Wille und Vorstellung*, ed. W. Frhr. von Löhneysen, 2 vols., (Frankfurt am Main, Suhrkamp 1986), 2: 590.

8 Cf. E. Vermeule, *Aspects of Death in Early Greek Art and Poetry,* (Berkeley/Los Angeles/ London, University of California Press 1979).

sophical connection, but rather in a mystical-religious, in particular Orphic-Pythagorean, context.[9]

Human mortality as a major philosophical issue presupposes the shift in focus of philosophical attention that in the course of the fifth century has cosmological thought replaced by anthropocentric thought. It is then that human behaviour in its moral aspects, various facets of epistemology, and other related issues are made the subject of philosophical discussion. In this connection a question naturally coming to the fore is what man really is, that is to say, man as a person. According to the Pythagorean doctrine of metempsychosis, the essential part of man is immortal *psuchè*, imprisoned in and therefore separable from the body.[10] Although an important contribution, this doctrine is still religiously inspired. It is Socrates, who is generally considered the main contributor of a philosophical concept of person.[11] For him, in the moral sphere as well as in the intellectual sphere, the essential part of man as a person, man's true 'self', is his *psuchè*. In consequence, philosophy should be ἐπιμέλεια τῆς ψυχῆς.[12] With death meaning dissolution, the question arises – being philosophically relevant in view of metaphysical and psychological implications – as to what is thought to happen to *psuchè* at death.

The various ways in which the problem of death is 'solved' appear to be linked to different views of *psuchè*, which in their turn are determined by a particular philosophy: obviously monistic materialism and transcendental metaphysics will employ widely different notions of *psuchè*. Wondering what death may be, Socrates proposes two possible answers in Plato's *Apology* (40c). These two possibilities may be said to be at the basis of all thoughts on the subject in Greek philosophy after him: "Death is one of two things. Either the dead man is as if he no longer exists, and has no sensations at all; or else as men say it is a change and migration of the soul from here to another place".[13] The former is worked out in various ways by those philosophers in whose conception everything, including *psuchè*, is reduced to matter, and by, albeit with some reser-

---

9 As to other 'strategies' "das Todesproblem zu entscharfen" cf. R. Rehn, "Tod und Unsterblichkeit in der platonischen Philosophie", in *Tod und Jenseits im Altertum*, edd. G. Binder-B. Effe, (Trier, Wissenschaftlicher Verlag 1991), 103–4; as to the *function* of religion and philosophy in respect of death cf. Schopenhauer, Die Welt, 2: 591.

10 Cf. W. Burkert, *Greek Religion,* (Oxford, Blackwell 1985), 300 and J. Barnes, *The Presocratic Philosophers,* 2nd ed., (Cambridge, Cambridge University Press 1982), 106 with note 2; as to the development of the concept of *psuchè* in general cf. D.B. Claus, *Toward the soul,* (New Haven/London, Yale University Press 1981) and J. Bremmer, *The Early Greek Concept of the Soul,* (Princeton, Princeton University Press 1983).

11 Concerning the legitimacy of this attribution originating with J. Burnet (The Socratic Doctrine of the Soul, *Proceedings of the British Academy* 7 [1915-6], 235–59) cf. Claus, *Soul*, 6–7, 156 sqq., 183; cf. also for a critical view G. Vlastos, *Socrates: Ironist and Moral Philosopher*, (Cambridge, Cambridge U.P.1991), 55 and note 37.

12 Cf. Plato, *Apol.* 29e and Burnet ad loc.; cf. also Plato, *Alcib.* 130c2-3: man is either nothing or nothing other than *psuchè*.

13 Transl. W.K.C. Guthrie, *A History of Greek Philosophy,* 6 vols., (Cambridge, Cambridge U.P. 1969), III: 479.

vation, Aristotle; the latter is worked out by upholders of philosophical dualism, of whom Plato is a major representative.

For reasons of space, it is evidently impossible to deal with the various elaborations in detail and throughout Antiquity. As for the first possibility I intend to confine myself to discussing the major views up to the end of the Hellenistic era (the end of the first century B.C.); as for the second possibility I shall confine my treatment to Plato. In this way no basic view will be left out. Moreover, in so far as modifications are introduced after the Hellenistic era, these do not amount to any basic changes.

If Socrates' first possibility is applied with retrospective effect to the few extant statements of Presocratic natural philosophers, then these seem to imply non-existence and, therefore, a complete lack of sensations. Unfortunately the evidence is meagre. However, as said above, it is unlikely for Presocratic philosophers to treat death as a philosophical subject in its own right. Statements on death in the context of natural philosophy have mainly physical import: at death *psuchè*, which broadly may be taken to stand for 'life-force', undergoes a physical change; which generally speaking will be an occurrence of merging into the particular stuff chosen by the natural philosopher in question as ἀρχή. Heraclitus' basic stuff is fire; consequently *psuchè* consists of fire and at death merges into fire, although this needs some qualification because Heraclitus may have distinguished between two types of death:

> One by fire for those cut down by violence in the prime of life and consciousness while still in possession of ψυχαί that are, at the instant of death, fiery; the other by water for those whose ψυχαί gradually become moist through disease.[14]

Evidently, such a view leaves no room for personal survival in any way.

The atomist Democritus, although technically not a *Pre*socratic, being a contemporary of Socrates, is a typical example of later Presocratic philosophy. In his view the whole of reality can be reduced to three basic principles: atoms, motion, and the void. The void is what separates the atoms and that through which they move. Everything, including *psuchè*, is to be explained in terms of these principles. *Psuchè* is conceived of as consisting of especially small round atoms. At death the individual *psuchè* is dispersed, and just as those of the body, its component particles are scattered throughout the universe. From extant fragments on 'ethics' (mainly maxims), whose authenticity is not undisputed, Democritus appears to have been strongly focused on the *Diesseits*. In view of his materialism it is to be expected that he thought it silly not to realize that death means "the dissolution of mortal nature" and to make oneself miserable by fears

14 Claus, *Soul,* 126; there are, however, some fragments that *might* be explained as implicating immortality of *psuchè*, viz. DK 62 (=47M[arcovich] =92K[ahn]) and DK 88 (=41M =93K).

of torments and punishments in an afterlife that does not exist anyway.[15] More than two centuries later his theory of atoms will serve as a physical basis of Epicurus' ethics.

Although Aristotle clearly does not belong to the tradition of monistic materialism, his view of death and its implications may legitimately be included in the present context of a discussion of death as non-existence. Initially, Aristotle seems to have endorsed Plato's dualism. He actually wrote a dialogue, the *Eudemus*, on the immortality of *psuchè*. Soon, however, he dissociated himself from dualism and gradually developed a concept of *psuchè* to fit in with his metaphysical theory of δύναμις and ἐνέργεια. The result finds suitable application in the field of biology, as is clear from his *De Anima*: *psuchè* is "the first actuality of a natural body having in it the capacity of life".[16] Man as a concrete individual is compounded of ὕλη (body) and εἶδος (*psuchè*). Death amounts to decomposition; it is disintegration into nothingness. This view is easily compatible with the almost 'clinical' approach of death implied in the following passage from *De Generatione Animalium*: "The process of formation, genesis, starts from not-being and advances till it reaches being; that of decay starts from being and goes back again till it reaches not-being".[17] Death is seen by Aristotle as a physical τέλος. It is, however, *not* a τέλος in the sense of an end aimed at, but in the sense of a terminal, the end of the line.[18] In the *Ethica Nicomachea*, when giving an example of objects of wish, Aristotle mentions immortality as a thing one can wish for, but an impossibility.[19]

Reservations as regards Aristotle's position arise from the following problem. In the *De Anima*[20] Aristotle distinguishes between νοῦς as νοῦς ποιητικός and νοῦς as νοῦς παθητικός.[21] It is in particular the status of the νοῦς ποιητικός that is problematic, for the reason that as Aristotle says: "It is this intellect which is separable and impassive and unmixed, being in its essential nature an activity" [a17-8], and

> It is, however, only when separated that it is its true self, and this, its essential nature, alone is *immortal and eternal*. But we do not remember, because this is impassive, while the intellect which can be affected is perishable and without this does not think at all. [a22-5, transl. Hicks; my italics]

---

15 DK 297; cf. Guthrie, *History,* (Cambridge 1965), II: 434 sqq.

16 *De Anima* II.i, 412a27; transl. Hicks

17 *De Gener. Anim.* II.v, 741b22; transl. Peck; cf. also *Parva Naturalia* 478b22-7

18 Cf. *Physica* II.ii, 194a27 sqq.

19 *E.N.* III.ii.7, 1111b22-3

20 III.v, 430a10 sqq.; cf. a.o. R.D. Hicks, *Aristotle De Anima,* ed. w. transl., introd. and notes, (Cambridge, Cambridge U.P. 1907 [repr. Amsterdam, Hakkert 1965]), LVIII sqq., esp. LXIV sqq.; and W.D. Ross, *Aristotle,* (London, Methuen 1971 [=1949^5]), 148 sqq.

21 This terminology is in part of a later origin; cf. Alexander Aphrod., *De Anima* 80.16-92.11, esp. 88.24-91.6, and id., *Mantissa* 106.19-113.24, esp. 107.29-110.3 B.

Νοῦς ποιητικός then, also said to have come into man from outside (θύραθεν), is immortal.[22] The problem is that what is εἶδος in the combined state of body-*psuchè* ceases to exist at death. So how is it possible for *part* of what ceases to exist to be itself immortal? The problem remains unsolved with Aristotle and indeed is unsolvable, because he offers no clues for its solution. What is clear, however, is that νοῦς ποιητικός cannot be personal in any relevant sense and that its immortality cannot possibly mean human personal survival.

When in *Ethica Nicomachea* I.x and xi Aristotle discusses the question whether the vicissitudes of their descendants can in any way reach the dead and affect their εὐδαιμονία, he is yielding to *Volksempfinden.* To think it out of the question is, he feels, "too unfriendly a doctrine... and one opposed to the opinions men hold".[23] As appears from *Topica* I.xi, 105a3 ff., a premiss that "deeply offends the religious and moral sentiments of men"[24] is not allowed. So it is understandable for Aristotle not to give short shrift to the opinion in question, which is at any rate not a suitable subject to be discussed dialectically.[25] Leaving aside the question itself, that is whether the dead really are in any way touched by what happens to their descendants, and for the moment assuming that somehow this is so, Aristotle expresses as his opinion that the effect produced will be too slight to affect the dead's εὐδαιμονία.[26] If, as seems to be the case, there is any effect, it will be of so little significance as to be negligible. Evidently Aristotle is working here in the context of a traditional view of the hereafter implying some form of future life.

In general it may be concluded that the idea of individual immortality in no way appeals to Aristotle: for man as a concrete individual death means the end. In his view, if it would be at all relevant to bring in the notion of immortality in such a context, it should be looked for in the continuity of the species:

> For it is the most natural function in all living things, if perfect and not defective or spontaneously generated, to reproduce their species; animal producing animal and plant plant, in order that they may, so far as they can, share in the eternal and the divine. For it is that which all things yearn after, and that is the final cause of their natural activity. ... Since, then, individual things are incapable of sharing continuously in the eternal and the divine, because nothing in the world of perishables can abide numerically one and the same, they partake in the eternal and divine, each in the only way it can, some more, some less. That is to say, each persists, though not in itself, yet in a representative which is specifically, not numerically, one with it.[27]

---

22 Cf. also *De Gener. Anim.* II.iii, 736b22–9.

23 *E.N.* I.xi.1, 1101a23–4.

24 Cf. J.A. Stewart, *Notes on the Nicomachean Ethics of Aristotle,* 2 vols., (Oxford, Oxford U.P. 1892), I: 139–40 and 149–50.

25 Cf. *Topica* I.x, 104a3 sqq.

26 *E.N.* I.xi.5, 1101b1–5.

27 *De Anima* II.iv, 415a26-b7; transl. Hicks; cf. also *Oec.* I.iii, 1343b23 sqq.

Epicurus was the first philosopher who set out to eliminate the fear of death by philosophical reasoning. The groundwork enabling him to develop his arguments was laid with his physics, which first and foremost served the practical purpose of explaining natural phenomena, because only clear insight into the workings of φύσις will free man from superstition and its ensuing fears. His physics teaches that the gods are completely extraneous to happenings in the world, which is neither their work nor governed by them, and that fear of possible punishment after death is futile, because *psuchè* does not in any way survive. All that occurs can be causally explained, even if it is impossible to establish the precise natural cause in every single instance. Basic to Epicurean physics are the two propositions that "nothing comes from nothing" and "nothing returns to nothing".[28] These two statements encapsulate the well-known principle *ex nihilo nihil fit,* applied already by the Ionian natural philosophers, although not then stated as a formula.[29] As in Democritean atomism reality is reduced to the three basic principles of atoms, motion, and the void. *Psuchè* according to Epicurus is "a corporeal thing, composed of fine particles, dispersed over all the frame".[30] (In the extant texts of Epicurus it is not *argued* that *psuchè* is a σῶμα, but Lucretius, III.161-76, adduces two reasons: 1) *psuchè* has the ability to move the body; the body being corporeal has to be moved by something that is itself corporeal ∴ *psuchè* is corporeal; 2) when the body suffers, *psuchè* suffers; pain can only be transmitted from body to body ∴ *psuchè* is corporeal.) All of the atoms of which *psuchè* is conceived to consist are smooth, fine and spherical.[31] As a result of this they tend to fall apart easily and have to be kept in place by the larger and more 'tightly assembled' atoms of the body. That is why on the destruction of the body the atoms of psuchè cannot endure as an independent 'sentient compound'. On the destruction of the body *psuchè* disintegrates into its component atoms. At the same time, however, the body, because sensations are due primarily to the atoms of *psuchè*, becomes insensitive. The disintegration of *psuchè* causes the body to lose its sensibility.[32] So there no longer is a subject of sensations.

Death, in a word, is ἀναισθησία. Lucretius, who argues more extensively against he fear of death than Epicurus seems to have done (at least judging by the extant remains),[33] likens this state of ἀναισθησία to the one before we were born:

> Death, then, is nought to us, nor does it concern us a whit, inasmuch as the nature of the mind is but a mortal possession. And even as in the time gone

28 Cf. Diogenes Laertius X.38–9.

29 Cf. Aristoteles, *Physica* I.iv, 187a32–5 and Lucretius, *De rerum natura* I.150 sqq.

30 Diogenes Laertius X.63, transl. Hicks; cf. also Lucretius, *D.R.N.* III.161–76, 177–230.

31 Cf. scholion Diogenes Laertius X.66 and Lucretius, *D.R.N.* III.86.

32 Cf. J.M. Rist, *Epicurus: An Introduction,* (Cambridge, Cambridge U.P. 1972), 80; as to a further nuance of the 'composition' of *psuchè* cf. a.o. Diogenes Laertius X.63 and Aetius, *Placita* IV.3.11 (p. 388 Diels); as to the distinction of a rational and an irrational 'part' cf. scholion Diogenes Laertius X.66; Rist, *Epicurus,* 79 sqq. and Lucretius, *D.R.N.* III.35 sqq. (distinction *animus/ anima*).

33 Cf. *D.R.N.* I.102–35; III.31–93 and 830-1094, esp. 830–42 and 973–7.

> by we felt no ill, when the Poeni came from all sides to the shock of the battle, when all the world, shaken by the tremorous turmoil of war, shuddered and reeled beneath the high coasts of heaven, in doubt to which people's sway must fall all human power by land and sea; so, when we shall be no more, when there shall have come the parting of body and soul, by whose union we are made one, you may know that nothing at all will be able to happen to us, who then will be no more, or stir our feeling; no, not if earth shall be mingled with sea, and sea with sky.[34]

If after death man is no more, what is there to be feared? This view has found expression in the pithy saying: ὁ θάνατος οὐδὲν πρὸς ἡμᾶς, death is nothing to us; while we exist, death is not with us, but when death comes, then we do not exist.[35] According to Epicurus, to live by philosophy it is essential to understand the nature of physical reality including the implications of death. The στοιχεῖα τοῦ καλῶς ζῆν – a right knowledge of the nature of the gods and a true understanding of death[36] – are expressly stated in the letter to Menoeceus, which in point of fact is an exposition of his moral philosophy for the general reader:

> Become accustomed to the belief that death is nothing to us. For all good and evil consists in sensation,[37] but death is deprivation of sensation. And therefore a right understanding that death is nothing to us makes the mortality of life enjoyable, not because it adds to it an infinite span of time, but because it takes away the craving for immortality. [125] For there is nothing terrible in life for the man who has truly comprehended that there is nothing terrible in not living. So that the man speaks but idly who says that he fears death, not because it will be painful when it comes, but because it is painful in anticipation. For that which gives us no trouble when it comes, is but an empty pain in anticipation. So death, the most terrifying of ills, is nothing to us, since so long as we exist, death is not with us, but when death comes, then we do not exist. It does not then concern either the living or the dead, since for the former it is not, and the latter are no more.[38]

Basically Epicurus' philosophy is hedonistic: the end, τέλος (εὐδαιμονία / μακαρίως ζῆν), is identified with pleasure, ἡδονή; the ultimate pleasure is freedom from pain in the body, ἀλυπία, and from trouble in the mind, ἀταραξία. So

34 Lucretius, *D.R.N.* III.830–42 (transl. Bailey); cf. Thomas Nagel, "Death", in Thomas Nagel, *Mortal Questions*, (Cambridge, Cambridge U.P. 1979), 7–8; cf. also E.M. Cioran, *De l'inconvénience d'être né,* (Paris, Gallimard 1973), 114.

35 Diogenes Laertius X.125; Epicurus, *Principal Doctrines* ii; Lucretius, *D.R.N.* III.830.

36 Diogenes Laertius X.123–7.

37 Cf. also Epicurus, *P.D.* ii.

38 Diogenes Laertius X.124–5 (transl. Bailey).

long as one is troubled by fear of death, ἀταραξία and therefore εὐδαιμονία is impossible.

Epicurean devotion to eliminating the fear of death was not shared by Stoicism, perhaps the most important of Hellenistic philosophies. Death, in their view, belonged to the *indifferentia*, that is to say, was considered morally indifferent. It is not surprising therefore to find that in early, 'orthodox', Stoicism little attention was paid to death and its implications.

The Stoics agreed with Epicurus in holding *psuchè* to be material; like Epicurean physics theirs was a materialist theory. As a clear understanding of the Stoic view on psuchè and the implications of death is impossible without some knowledge of their physics, a brief description of its main tenets has to be included.

The Stoics believe that there is one single world order or cosmos, which they explain from two basic principles, the active, τὸ ποιοῦν, and the passive, τὸ πάσχον. The passive principle is indeterminate matter, or substance without qualitative determination, ὑλή or ἄποιος οὐσία; the active is a formative principle; conceived of as material, it is finer; it is also called *Logos* or God. There is no trace whatsoever of dualism: the two principles are inseparable, God and matter always being conjoined. The world order is the whole, τὸ ὅλον, but not τὸ πᾶν, all that is, the latter also embracing the void. Conceived of materially the active principle is fire and described in more specific terms as πνεῦμα ἔνθερμον or πῦρ τεχνικόν. Penetrating everything this fiery πνεῦμα, in maintaining a constant tension, operates as a cohesive force. In the initial condition this pure fire is all there is; then, through an intermediate stage of respectively air and water, the four elements are produced – earth, water, air and (destructive) fire –, which phase is the beginning of the formation of the cosmos,[39] or rather the origin of *a* cosmos. For it is believed that in the ongoing process of conversion fire gradually becomes the dominant element; that as a result of this, after a certain period of existence, the cosmos is finally destroyed in an all-consuming fire, the so-called ἐκπύρωσις, and that following the conflagration the cosmogonic process repeats itself. A strange feature is the accompanying belief in the eternally repeated sequence of events. In an everlasting recurrence the same things are happening again and again, that is to say, every new world means an exact recurrence of everything that happened in the preceding one: Socrates drinking hemlock is an ever-recurrent event.[40]

Logos pervades the world. All things have in them a fragment, ἀπόσπασμα, of the divine *Logos*, which as a seminal principle, λόγος σπερματικός,

---

39 As to Stoic cosmogony cf. D.E. Hahm, *The Origins of Stoic Cosmology,* (Columbus, Ohio State U.P. 1977), 57 sqq.

40 Such a view originates from the thought that God (= the active principle) does everything in the best possible way: if in a new cosmos something should happen in another way, that might implicate that this same thing had not been realized by God in the best possible way in the previous cosmos; cf. M. Pohlenz, *Die Stoa,* 2nd-3rd ed., 2 vols., (Göttingen, Vandenhoeck & Ruprecht 1959–64), I: 79 sqq.; cf. also A.A. Long-D.N. Sedley, *The Hellenistic Philosophers,* 2 vols., (Cambridge, Cambridge U.P. 1987), I: 311.

defines their being. It is, as it were, a thing's natural blueprint. The λόγοι σπερματικοί reveal themselves in various ways: in living things they operate as ψυχή, and in a man in his fully mature state as ψυχὴ λογική or λόγος. *Psuchè* then is material. Although sometimes referred to as 'fire', it is more often described as τὸ συμφυὲς ἡμῖν πνεῦμα or πνεῦμα σύμφυτον ἡμῖν συνεχὲς πάντι τῷ σώματι διῆκον.[41] Πνεῦμα in this context may be found designated as νοερὸν θερμόν.[42] *Psuchè* has its origin in the process of reproduction, the procreator's *psuchè* imparting πνεῦμα at the moment of conception. This πνεῦμα is nourished by exhalation of the blood and in the course of time grows into a fully developed human *psuchè*.[43] In view of this materialism it is to be expected that immortality of *psuchè* has no place in orthodox Stoicism. Death is disintegration of the composite made up of body and (material) *psuchè*. What does not decompose is πνεῦμα. *Psuchè*'s πνεῦμα remains after death, but it was debated how long. According to Zeno *psuchè*, πολυχρόνιον πνεῦμα, would gradually lose its tension and merge into the All.[44] Cleanthes believed all *psuchai* to survive until the ἐκπύρωσις,[45] whereas Chrysippus held the view that only the *psuchè* of the wise man was to survive until then.[46] However, what is at issue here, is clearly not personal survival. In another διακόσμησις every *psuchè* comes into being again. This may seem absurd, but it is no stranger than, and in fact a corollary of, the aforementioned everlasting recurrence of the same things.[47]
Having perished by merging into the primaeval fire the *psuchai* are newly generated. This alone would make personal immortality, if any, impossible, because continuous survival is thereby precluded.

The Stoics then could see no reason to concern themselves with the fear of death. Firstly, because an eschatological view of the beyond is ruled out by their physics, and secondly because, as was said earlier, death is one of the things that are morally indifferent, that is to say, one of the things intermediate between good and bad. Although within the category of *indifferentia* a distinction was made between things 'preferable', προηγμένα, and things 'dislikeable', ἀποπροηγμένα, inviting a 'preferably not', this was of no consequence to their moral status. Examples of things to be preferred are wealth and health, examples of things inviting a 'preferably not' are poverty and illness. Another example of the latter is death. Now theory is one thing, practice quite another matter. It does not seem at all impossible to have a clear understanding of Stoic physics and yet be disturbed by fear of death. To the Stoic sage, however, this is inconceivable, for he is ἀπαθής. What exactly this means, can only be understood correctly in connection

41 J. von Arnim, *Stoicorum Veterum Fragmenta (= SVF)*, 4 vols., (Stuttgart, Teubner 1968 [=Leipzig 1903–24]), I.134: 37–8; II. 774,778: 217; II.885: 238.
42 *SVF* II.779: 217.
43 Cf. Pohlenz, *Die Stoa*, I: 85 sqq.
44 *SVF*, I.146: 40.
45 *SVF*, I.522: 118.
46 *SVF*, I.522: 118 and II.811: 223.
47 *SVF*, II.623–4: 189–90.

with the Stoics' peculiar notion of πάθος. According to the Stoics a πάθος, such as fear, is a ὁρμὴ πλεονάζουσα, i.e. an impulse that is getting out of hand, an excessive uncontrolled 'drive'. It is an instance of disturbance due to misjudging caused by an overestimation of indifferent things. Nothing of the kind ever happens to the Stoic sage. He knows his place in the cosmic order and accepts with equanimity all that happens to him as providentially ordered: "ducunt volentem fata, nolentem trahunt".[48] Only when circumstances, e.g. through fatal illness, make 'living in accordance with Nature', the Stoic τέλος of life, no longer possible, only then is it permissable to take one's own life. To depart this life must be the reasonable thing to do, hence εὔλογος ἐξαγωγή.[49]

The Stoic sage is unexampled, something the Stoics themselves recognised. They were well aware of the enormous gap between their ideal and everyday life. The majority of people, although born with a tendency towards the good, were, of course, far from identifying 'living in accordance with Nature' with εὐδαιμονία, due to, among other things, the corrupting influence of their environment.[50] It goes without saying that by Stoic standards all those belonging to the φαῦλοι, that is the 'unwise', lacked ἀπάθεια; no doubt, they all invariably dreaded death.

Later Stoicism is seen to introduce modifications in these, as well as in other aspects. It is also noticeably influenced by Platonism, which in particular colours the view of death and its implications. These later developments, however, are outside the scope of this article.[51]

The second of the possibilities mentioned by Socrates in Plato's *Apology* (40c: "or else as men say it is a change and migration of the *psuchè* from here to another place") derives from a religious rather than philosophic outlook, viz. the Orphic-Pythagorean conviction that man's *psuchè* is immortal and passes through a series of incarnations in a cyclical pattern. This conviction presupposes dualism: the belief that body and *psuchè* are two entirely different entities.[52] *Psuchè* enters the body from elsewhere and appears to be identified with man's self, the person. In the Orphics' view, being incarnated in a new body is a kind of punishment (σῶμα σῆμα). By living as best he can, man should secure a progressively better reincarnation, that is to say, try to avoid any lower form of life, in order to attain complete purification of his *psuchè*, which from then on, liberated from incarnation, will enjoy eternal bliss. Death, in such a view, may be looked forward to as a final release.[53] It is the Orphic-Pythagorean tradition that Socrates

48 Seneca, *Epistulae Morales* 107.11.

49 Cf. *SVF*, III.757 sqq.: 187 sqq.

50 Cf. *SVF*, III.228–36: 53–6.

51 As to this development cf. Pohlenz, *Die Stoa*, I: 229 sqq. and II: 115–6.

52 For a clear wording of such a dualistic view in a Platonic context cf. *Gorgias* 524b2–4.

53 The question is, of course, what after all is reincarnated: is the human personality preserved and is this reincarnated every time anew? The material we have to our disposal is not sufficient to enable us to answer this question with certainty.

has in mind when in Plato's *Meno* he introduces the theory of Recollection, ἀνάμνησις.[54] The purpose of bringing this in is to evade the seemingly incontestable argument against the acquisition of new knowledge, which runs as follows: searching for knowledge and learning is impossible, because on the one hand there is no point in seeking or learning what one knows already, and on the other hand no man can seek to find out or learn what he does not know, for this implies knowing what he is looking for, or having the means of recognising it when found. The doctrine of Recollection is expounded by certain priests and priestesses as well as by "divinely inspired" poets such as Pindar:

> they say that the soul of man is immortal, and at one time comes to an end, which is called dying, and at another is born again, but never perishes. ... Seeing then that the soul is immortal and has been born many times, and has beheld all things both in this world and in the nether realms, she has acquired knowledge of all and everything; so that it is no wonder that she should be able to recollect all that she knew before about virtue and other things. For as all nature is akin, and the soul has learned all things, there is no reason why we should not, by remembering one single thing – an act which men call learning – discover everything else, if we have courage and faint not in the search; since it would seem, research and learning are wholly recollection.[55]

In the *Meno*, the doctrine is invoked in order to enable Socrates to make the statement that learning is ἀνάμνησις, *re*-collection of knowledge acquired in previous incarnations: if learning is a form of remembering, then it is possible for Socrates to say that he is not learning anything new, but only bringing to mind what he knew already. *Psuchè* is depicted as the knowing subject: pre-existing, it passed through a number of incarnations, in which it acquired knowledge; in other words, it does not depend for its knowledge on its present existence, but its knowledge comes from an independent source. The nature of its knowledge is left unspecified, that is to say, nothing is said to indicate what kind of things the objects of knowledge are imagined to be. The theory is illustrated in the *Meno* by having a slave solve a geometrical problem; that, though ignorant of mathematics, he is able to do so must be due to ἀνάμνησις.[56]

The immortality of *psuchè* is at the centre of interest of Plato's *Phaedo*, whose dramatic setting is the last day of Socrates' life. In the course of the dialogue several 'proofs'[57] of immortality are developed, although no attempt is made to examine the notion of immortality itself; it seems to be taken for granted

54 *Meno* 81a5 sqq.

55 *Meno* 81b3–5 and c5–d5 (transl. W.R.M. Lamb).

56 *Meno* 82b9 sqq.

57 *Phaedo* 69e–85b; cf. J.E. Raven, *Plato's Thought in the Making*, Cambridge, Cambridge U.P. 1965), 81 sqq. and his 'evaluation' on p. 103: "... the fallacies on which all the Phaedo's quasi-scientific arguments for immortality rest."

that ἀθάνατος implies 'survival through time' or 'perdurance through time'.[58] The first of the proofs is the so-called 'argument from opposites':

> 'Let us consider in every case whether things are not all born simply from their opposites – everything, that is, which *has* an opposite, in the way in which the beautiful is opposite to the ugly, and just to unjust, and so on. Let us consider, I say, whether all things which have an opposite are not born simply from their opposites. For example, when something "larger" comes into being, it must, I suppose, become "larger" after having previously been "smaller". 'Yes.' 'Then similarly if it becomes "smaller", it will only become "smaller" after having formerly been "larger"?' 'Yes,' he said. [70e]...'Then we are satisfied', he said, 'that in every instance generation is of opposites *from* opposites?' 'Yes.' 'Well, perhaps we may say something of this sort about them, that between every pair of opposites there are two processes of 'coming into being' – from the one into the other, and then from the other back again into the former. ... So separating and combining, growing cold and growing hot, and all such processes ... must ... in fact always be in the same case: they must always arise out of each other, and there must always be a process of generation from the one into the other' [71a,b].[59]

This is then applied to the opposites 'living' and 'being dead': the two states arise out of each other. The dead arises out of the living, and as Socrates' interlocutor cannot but admit, the living from the dead [71d-e]. Socrates continues as follows:

> 'Then all living men and all living things are born from the dead?' 'It appears so,' he (*viz. Cebes*) said. 'Then our souls,' he said, 'exist in Hades.' 'It seems so.' 'Well, one of the two processes concerned is also clear, surely? Dying is clear enough, I presume?' 'Certainly,' he (*viz. Cebes*) said. 'What shall we do then?' he said. 'Shall we not posit, for the sake of balance, the existence of the opposite process, or will nature be lame in this respect? Surely we must allow that there is a process opposite to dying?' 'Most certainly,' he (*viz. Cebes*) said. 'What is it then?' 'Coming to life again.' 'Then,' he (*viz. Socrates*) said, 'if there is such a thing as coming to life again, this "coming to life again" will be a process of generation by which one leaves the dead to join the living?' 'Yes.' 'We are agreed then, that in this way the living have been born from the dead

58 These terms are borrowed from Charles L. Griswold, *Self-Knowledge in Plato's Phaedrus*, (New Haven/London, Yale U.P. 1986), 145; as to some problematic aspects of Plato's notion of immortality cf. A. Flew, "Immortality", in *The Encyclopedia of Philosophy*, ed. P. Edwards, 8 vols., (New York/London, Collier-MacMillan 1972 [=1967]), IV:139 sqq.

59 *Phaedo* 70e-71b; all passages from the Phaedo are cited in Bluck's translation [R.S. Bluck, *Plato's Phaedo*, a Translation with Introd., Notes and Appendices, (London, Routledge/Kegan Paul 1955)].

> no less than the dead have been "born" from the living? And we decided, I think, that if this is true [70d: "...if it really could be made plain that the living are born from the dead and from nowhere else..."] there is sufficient evidence that the souls of the dead must exist somewhere – in some place, I mean, from which to be born again'.[60] ...'I think, Cebes, ... that in actual fact there is such a thing as coming to life again as well, and that the living are born from the dead, and that the souls of the dead have real existence'.[61]

The attempt to prove the Orphic doctrine of the 'cycle', for which reason the argument is also referred to as the 'cyclical' argument, is immediately followed by the argument from recollection. The *Phaedo*, unlike the *Meno*, is specific in its description of the nature of the knowledge recovered: the objects of the knowledge that is brought to mind through ἀνάμνησις are the Forms. *Psuchè* has this knowledge 'a priori',[62] that is to say, from long before, from a time when it must have existed on its own, apart from body, and so was not provided with data by those bodily instruments the senses; the only things for *psuchè* to become acquainted with then, the sole things to 'see', were the Forms. In connection with this Plato advances his 'proof' from recollection, which is illustrated by the following example: no two sensible things, e.g. sticks or stones, are ever equal; yet seeing two sticks that are approximately equal reminds us of that which really *is* equal. The explanation for this must be that we have been acquainted with perfect equality before we were born. So we must have existed earlier on and we must have had intelligence then.[63] It appears that the dualism of body-*psuchè* found in the passage quoted from the *Meno* is coupled with a dualism about knowledge: the knowledge whose objects are the Forms has *psuchè* for its subject; the knowledge whose objects are sensible things has for its subject the body, or rather the senses. The so-called argument from affinity, a third 'proof' meeting the objection how we know that *psuchè* is not dispersed at death, brings in ontological dualism: there are two kinds of entities, viz. sensible things and Forms:

> 'Do you want us then,' he (*viz. Socrates*) said, 'to lay it down that there are two kinds of existents, the one kind seen, the other invisible?' 'Let us do that,' he (*viz. Cebes*) said. 'And that the invisible kind is always constant, while the seen is never constant?' 'That too,' he (*viz. Cebes*) said.[64]

This ontological dualism goes hand in hand with epistemological dualism. There are two different states of knowing: having the Forms for its objects, *psuchè*'s state is ἐπιστήμη; having the αἰσθητά for its objects, the senses' state of knowing

---

60 *Phaedo* 71d-72a.

61 *Phaedo* 72d-e.

62 ἐν προτέρῳ τινὶ χρόνῳ, *Phaedo* 72e.

63 *Phaedo* 74a-76d.

64 *Phaedo* 79a6-11.

is δόξα. The fact of having the Forms as objects of knowledge implies that *psuchè* is immortal. This is in line with the principle known from Presocratic 'theory of knowledge' *similia similibus (cognoscuntur)*: the Forms are imperishable, therefore the subject knowing them must be imperishable. There is difficulty in that, strictly speaking, *psuchè* itself would have to be a Form; to the extent that *psuchè* is not a Form, but *akin* (συγγενής) to Form, Plato is inconsistent.[65]

> 'Again, we said this some time ago, that when the soul makes use of the assistance of the body for the study of something, by using sight or hearing or some other sense – for this is the bodily method, the study of something through sense – it is dragged by the body towards what is never constant, and it vacillates, and is confused, and dithers as though it were drunk, because it is in contact with things that are in that sort of state?' 'Yes.' 'But when it makes enquiry all by itself, it goes off there to what is pure and everlasting and deathless and invariable, and as though akin to it always remains with that kind of being, whenever it is by itself and can do so; it ceases to vacillate and is always constant and invariable in its relationship with those realities, because it is in contact with things that are themselves constant and invariable. And this condition of it is called contemplation.[66] Isn't that so?'[67]

The latter, clearly, is the better of the two conditions. It is achieved when, unrestricted by a body liable to dissolution, i.e. not incarnated, *psuchè* 'sees' the Forms *by itself*, that is to say, after death. Not every *psuchè*, however, will attain this, but only the *psuchai* of those who while living have 'separated' themselves from their bodies so far as they can, the philosophers. So when he has Socrates reply: "ὅτι τῷ ὄντι οἱ φιλοσοφοῦντες θαναθῶσι",[68] Plato means to bring out that those who are true philosophers choose to live their lives shunning their bodies, striving to become independent of these, so as to free their *psuchai* and create optimal conditions for abstract reasoning to a degree as is distinctive of *psuchè*'s state after death. The *Phaedo* does not elaborate on the cycle of reincarnation, but in the *Phaedrus* Plato returns to the subject. The latter dialogue develops yet another 'proof' of *psuchè*'s immortality,[69] in the context of which it becomes clear that Plato fails to distinguish between two concepts of *psuchè*: in addition to its being identified with the person, the bearer of moral and intellectual values,

---

65 The problem that arises from this, viz. how one can know at all that something is a *psuchè*, Plato does not solve.

66 The Greek reads here Φρόνησις which would be rendered more accurately by 'insight' or 'wisdom'.

67 *Phaedo* 79c2–d7.

68 "that philosophers are, in very fact, more dead than alive"; *Phaedo* 64b4–5.

69 For yet another, if rather unconvincing, proof cf. *Resp.* X.608c sqq.; cf. also Guthrie, *History*, IV: 555.

psuchè is now made the principle of motion, its own motion as well as that of the body.[70]

> 'All soul is immortal; for that which is ever in motion is immortal. But that which while imparting motion is itself moved by something else can cease to be in motion, and therefore can cease to live; it is only that which moves itself that never intermits its motion, inasmuch as it cannot abandon its own nature; moreover this self-mover is the source and first principle of motion for all other things that are moved. Now a first principle cannot come into being: for while anything that comes to be must come to be from a first principle, the latter itself cannot come to be from anything whatsoever: if it did, it would cease any longer to be a first principle. Furthermore, since it does not come into being, it must be imperishable: for assuredly if a first principle were to be destroyed, nothing could come to be out of it, nor could anything bring the principle itself back into existence, seeing that a first principle is needed for anything to come into being. The self-mover, then, is the first principle of motion: and it is as impossible that it should be destroyed as that it should come into being: were it otherwise, the whole universe, the whole of that which comes to be, would collapse into immobility, and never find another source of motion to bring it back into being. And now that we have seen that that which is moved by itself is immortal, we shall feel no scruple in affirming that precisely that is the essence and definition of soul, to wit self-motion. Any body that has an external source of motion is soulless; but a body deriving its motion from a source within itself is animate or *besouled*, which implies that the nature of soul is what has been said. And if this last assertion is correct, namely that 'that which moves itself' is precisely identifiable with soul, it must follow that soul is not born and does not die'.[71]

In the first sentence ψυχὴ πᾶσα ἀθάνατος presents a difficulty. Does ψυχὴ πᾶσα mean 'every *psuchè* ' or 'all *psuchè* 'or '*psuchè* in all its forms' or '*psuchè* in its totality'?[72] Taking it to mean '*psuchè* in its totality' seems incompatible with the view of *psuchè* as introduced in the *Politeia*. In this dialogue Plato distinguishes three 'parts' within *psuchè*, three ways of functioning of *psuchè* : the λογιστικόν

70 As to this double concept of *psuchè*, i.e. a philosophical concept and a concept of a rather mystical-religious origin which Plato sometimes seems to identify, cf. I.M. Crombie, *An Examination of Plato's Doctrines*, 3rd ed., 2 vols., (London, Routledge/Kegan Paul 1969-1971), I: 316 sqq.; Guthrie, *History*, IV: 348.

71 *Phaedrus* 245c–246a; all passages from the Phaedrus are cited in Hackforth's translation [R. Hackforth, *Plato's Phaedrus*, Translated with Introd. and Comment., 2nd ed., (Cambridge, Cambridge U.P. 1952)]; for an analysis of the argumentation (a reconstruction in the form of a sorites) cf. Griswold, *Self-Knowledge*, 82.

72 For a discussion of this problem cf. Hackforth, *Plato's Phaedrus*, 64; Guthrie, *History*, IV: 419 with note 4; Griswold, *Self-Knowledge*, 84.

(the reasoning 'part'), the θυμοειδές (the impulsive 'part', 'spirit') and the ἐπιθυμητικόν (the appetitive 'part').[73] Immortality is now confined to the λογιστικόν, although it is hard to see what we are to make of this: a surviving function of abstract reasoning? However, it is difficult to conceive of the ἐπιθυμητικόν, which is inevitably bound up with the body, as being immortal. Tripartite *psuchè* as introduced in the *Politeia* is made use of in the *Phaedrus* in a passage immediately following the argument for immortality. Triplicity here is *not* confined to *psuchè*'s incarnate state; even the *psuchai* of the gods are tripartite.[74] In the passage in question Plato describes the nature of *psuchè* in a myth of great vision, but with a certain vagueness of detail.[75] Tripartite *psuchè* is compared to a team of winged horses driven by a winged charioteer. The horses belonging to the *psuchai* of the gods are all good, but the human *psuchè* has one good horse and one bad. The first part of the myth describes how a procession of *psuchai*, headed by those of the gods, proceeds along the rim of the οὐρανός. There is said to be a steep ascent leading to the top of the arch supporting the οὐρανός. Climbing the ascent is easy for the *psuchai* of the gods, but not for the other *psuchai*, for the bad horse through its weight is pulling the human *psuchè* down towards the earth [247a-b]. The *psuchai* that are called immortal, when they are at the top: "come forth and stand upon the back of the world: and straightway the revolving heaven carries them round, and they look upon the regions without".[76] The immediately following explains what is meant by the "regions without"; it is the place where: "true Being dwells, without colour or shape, that cannot be touched, reason alone, the soul's pilot, can behold it, and all true knowledge is knowledge thereof";[77] that is to say, the charioteer (=λογιστικόν) has come in view of the world of the Forms, which, to Plato, is what really and truly *is*. It is what the *psuchai* of the gods unfailingly experience: while being carried round on the revolving οὐρανός the divine *psuchè* is nourished by this sight and fully satisfied comes back to its starting point. The other *psuchai* fare less well:

> 'of the other souls that which best follows a god and becomes most like thereunto raises her charioteer's head into the outer region, and is carried round with the gods in the revolution, but being confounded by her steeds she has much ado to discern the things that are; another now rises, and now sinks, and by reason of her unruly steeds sees in part, but in part sees not. As for the rest, though all are eager to reach the heights and seek to follow, they are not able: sucked down as they travel they trample and tread upon one another, this one striving to outstrip that. Thus confusion

73 *Resp.* IV.434d–441c.

74 For a suggestion to solve this problem cf. Guthrie, *History*, IV: 442 sqq.

75 *Phaedrus* 246a sqq.; one should, of course, appreciate the function of myth with Plato: he tries to express, often in allegorical and/or poetical form, what rational argument cannot convey; for a critical view of this traditional interpretation of the role of myth with Plato cf. Griswold, *Self-Knowledge*, 142 sqq.

76 *Phaedrus* 247b–c.

77 *Phaedrus* 247c.

> ensues, and conflict and grievous sweat: whereupon, with their charioteers powerless, many are lamed, and many have their wings all broken; and for all their toiling they are baulked, every one, of the full vision of Being, and departing therefrom, they feed upon the food of semblance'.[78]

In this connection Plato comes back to reincarnation and he goes on to describe how he imagines the cycle to be:

> 'Hear now the ordinance of Necessity. Whatsoever soul has followed in the train of a god, and discerned something of truth, shall be kept from sorrow until a new revolution shall begin; and if she can do this always, she shall remain always free from hurt. But when she is not able so to follow, and sees none of it, but meeting with some mischance comes to be burdened with a load of forgetfulness and wrongdoing, and because of that burden sheds her wings and falls to the earth, then thus runs the law: in her first birth she shall not be planted in any brute beast, but the soul that hath seen the most of Being shall enter into the human babe that shall grow into a seeker after wisdom or beauty, a follower of the Muses and a lover; the next, having seen less, shall dwell in a king that abides by law, or a warrior and ruler; the third in a statesman, a man of business or a trader; the fourth in an athlete or physical trainer or physician; the fifth shall have the life of a prophet or a mystery-priest; to the sixth that of a poet or other imitative artist shall be fittingly given; the seventh shall live in an artisan or farmer; the eighth in a sophist or demagogue, the ninth in a tyrant'.[79]

In general it takes 10,000 years before *psuchè* has regrown its wings and may return to its heavenly home. The philosopher, however, is in an exceptional position: in his case it takes 3000 years. After every thousand years there is to be a new reincarnation, partly determined by lot, partly by choice. The period between a life and a new incarnation is the time to be punished or rewarded. The first incarnation will be in a human body; subsequent incarnations may be in animal bodies.[80] The philosopher's cycle is completed sooner because his life is dominated by a constant devotion to the Forms;[81] in terms of the myth: only the philosopher's *psuchè* regrows its wings.

Actually, immortality as described in the *Phaedrus* is no different from immortality as described in the *Politeia*: the λογιστικόν alone 'sees' the Forms, that is to say, only the rational 'part' of *psuchè* is immortal. It is therefore subject

---

78 *Phaedrus* 248a–b.

79 *Phaedrus* 248c–e; on the fate of *psuchè* after death, the way in which the reincarnations come about included, cf. also the myth of Er, *Resp.* X.613e sqq., and for a much simpler and shorter myth on the judgement of the *psuchai* after death cf. *Gorgias* 523a sqq.; cf. also *Phaedo* 107a sqq.

80 This, however, is contested; cf. Hackforth, *Plato's Phaedrus*, 88; on the transformations into animals cf. also *Timaeus* 91d sqq.

81 Cf. the θανατᾶν of the philosopher in the Phaedo!

to the same criticism: it does not seem to come anywhere near to *personal* survival or immortality.[82]

Plato's theory of immortality is an integral part of his metaphysics. It is not devised with the intention of eliminating the fear of death, but for anyone embracing the theory to fear death would certainly be inappropriate. As has been pointed out by M. Jager, in the context of Plato's philosophy fear of death must be attributed to calling in question the indestructibility of *psuchè*; it will be removed by resolving these doubts through rational argument showing the belief in immortality to be firmly based.[83] There is, it would appear, a moral aspect to the question of immortality for Plato. Socrates' comment to the myth of Er at the end of the *Politeia* seems to imply that it is morally incumbent on man to believe that one day, after death, he will have to account for his conduct throughout life:[84]

> 'And so, Glaucon, the tale (*scil. of Er*) was saved from perishing; and if we will listen, it may save us, and all will be well when we cross the river of Lethe. Also we shall not defile our souls; but, if you will believe with me that the soul is immortal and able to endure all good and ill, we shall keep always to the upward way and in all things pursue justice with the help of wisdom. Then we shall be at peace with Heaven and with ourselves, both during our sejourn here and when, like victors in the Games collecting gifts from their friends, we receive the price of justice; and so, not here only, but in the journey of a thousand years of which I have told you, we shall fare well'.[85]

The theories of the philosophers concerning the implications of death were probably in their original context mainly known to a limited, select circle of intellectuals. In a wider context, therefore, the effect of their theories, whether they were intended to eliminate the fear of death or not, will have been slight. To conclude, therefore, I will pay some attention to a philosophical-literary genre that apparently did reach a wider public and that was explicitly meant to console people in the event of a bereavement. It is a genre known as consolation-literature, whose spiritual father was the Academic Crantor (±335-275 B.C.), by his *Peri penthous*, a work characterized by H.-Th. Johann as a "psychotherapeutischer Bestseller der Antike".[86] To achieve the objective – consolation –

82 For a non-literal explanation of 'immortality' in the myth, in the sense of "the capacity of a mind that exists in time to think what is eternal", cf. Griswold, *Self-Knowledge*, 145.

83 M. Jager, "Antieke filosofen over de dood", *Lampas* XIII.4 (1980), 308.

84 Thus Rehn, Tod, 113; cf. also Rehns remark (ibid., 114) "Der Gedanke an ein Jenseits, in dem die Guten belohnt und die Bösen bestraft werden, ist nur sinnvoll, wenn von einer individuellen Unsterblichkeit ausgegangen wird".

85 *Resp.* X.621b8-d3; transl. F.M. Cornford, *The Republic of Plato*, (London/Oxford/New York, Oxford U.P. 1974 [=1941]).

86 H.- Th. Johann, *Trauer und Trost; eine quellen- und strukturanalytische Untersuchung der philosophischen Trostschriften über den Tod*, (München, W. Fink Verlag 1968), 13; on the genre cf. R. Kassel, *Untersuchungen zur griechischen und römischen Konsolationsliteratur*, Zetemata 18, (München, C.H. Beck Verlag 1958), und Johann, *Trauer.*

arguments from different philosophical origins are used. Epicurean arguments in particular are used to show that there will be nothing after death and thus that nothing has to be feared, and Platonic arguments to advocate death as a means of transition to another (and better) state. We find a similar approach in Seneca's *Consolationes* among other works, in Plutarch's *Consolatio ad Apollonium*, and in really quite a clumsy example, that nevertheless well illustrates the principles of this approach, viz. the pseudo-Platonic dialogue, *Axiochus*. The problem of death is central in this little dialogue, which presumably dates from the first century B.C.[87] and which especially during the Renaissance enjoyed a popularity[88] that for us in retrospect seems hard to understand. Its sub-title is accordingly "or On death". This problem is not treated in the manner of a 'dogmatic' treatise where one specific position is taken and differing views are contested. Rather, we find in this little work an argumentation for the most part along Epicurean and Platonic lines; in addition to this, but on a much smaller scale, there are some elements taken from Cynic and Stoic sources.

As stated above, the juxtaposition of such, often incompatible, heterogeneous elements taken from different philosophical systems is a common phenomenon in the texts belonging to the category of consolation-literature. As regards the phenomenon that Epicurean and Platonic arguments are found side by side in one and the same consolatory text, it is in all probability Plato, *Apology* 40c, which was the archetypal example.[89] Epicurean argumentation was after all very convenient in developing the first possibility mentioned in the passage from the *Apology*, as H.- Th. Johann also points out: "für die Ausmalung der Nichtexistenz und der absoluten Empfindungslosigkeit im Tode (boten sich) die epikureischen Farben jedem *consolator* ungeachtet seiner Grundeinstellung zur Lehre des Kepos [*i.e. the school of Epicurus*] an",[90] while, of course, Plato himself could serve for developing the second possibility.

At the beginning of the dialogue Axiochus' son appeals to Socrates to come and console his father who is dying and who is frightened of death; the purpose which must justify all subsequent argumentation is formulated in 364c2: "that he may meet his fate without complaint".[91]

Next follows what may be considered as some encouragement for those who are faced with death or the fear of death: "life is a brief stay in a foreign land .... it's necessary for those who spent it reasonably well to meet their destiny cheerfully,

---

87 Cf. J. Chevalier, *Étude critique du dialogue pseudo-platonicien l'AXIOCHUS sur la mort et sur l'immortalité de l'âme*, (Paris, F. Alcan 1914), 115; cf. also Platon, *Dialogues apocryphes*, ed. J. Souilhé, Oeuvres complètes XIII.3, (Paris, Les Belles Lettres 1962), 135, and J.P. Hershbell, *Pseudo-Plato, Axiochus*, Texts and Translations 21/Graeco-Roman Religion Series 6, (Chico, Scholars Press 1981), 20-1.

88 cf. Chevalier, Étude, 117 ("la Renaissance lut, traduisit et commenta l'*Axiochos* avec passion. C'est à travers ce dialogue apocryphe que les humanistes découvrirent Platon.")

89 See above p. 198.

90 *Trauer*, 124 (§261), following Kassel, *Untersuchungen*, 79.

91 All passages from the Axiochus are cited in the translation by Hershbell; as to the Stoic term 'fate' (τὸ χρεών) cf. *SVF* II. 914:265.

all but singing a paean of praise." Axiochus' fear, however, is not removed by this. Axiochus appears to be afraid of what will happen to his body after death or of the perception of this [365c1-7]. This fear is disposed of with Epicurean arguments:

> SOCR. 'But, Axiochus, because of your thoughtlessness, you uncritically connect sensation with absence of sensation; and you are doing and saying things contrary to yourself, not realizing that at one and the same time you lament the absence of sensation and are pained at decay and loss of pleasures, just as if by dying you entered into another life instead of having lapsed into complete insensibility such as you had before birth. So, for example, just as in the administration of Draco and Cleisthenes there was nothing evil that concerned you – for it is elementary that you, whom the evil could have concerned, did not exist – so not even after death there will be any evil. For you, whom it would concern, will not exist.' [365d1-e1]

Then follows in 365e2 an abrupt transition from Epicurean argumentation to one that is, for the most part, Platonic:

> 'Away, then, with all this nonsense, and realize this: that once the union of body and soul is dissolved and the soul has been established in its proper place, the corpse which remains, being earthly and irrational, is not the human person. For we are soul, an immortal living being, locked up in a mortal prison.' [365e2-366a1]

Perception of what will happen to the body after death now being eliminated and the body no longer playing a role, Socrates can pass on to *psuchè*. This constitutes man proper and is immortal. Next Socrates characterizes the body as earthly and irrational, as a mortal prison of *psuchè*, and as a tent for the suffering of evil. In this way he emphasizes the contrast between the body which does not constitute the human person, and *psuchè* which is the essence of human person and an immortal living being. Thus death means liberating *psuchè* from its mortal prison, enabling it to return to its "own place".[92] So the release from this life which death implies is, Socrates concludes, a "change from a kind of evil to a good".[93] In fact, between 365b1 and 366b1, in a comparatively small compass, all is said. The two possibilities Socrates considers in Plato, *Apology* 40c, have been brought to the fore, the first one in Epicurean, the second one in Platonic wording.[94] Axiochus reacts sceptically to Socrates' pronouncing death to be a change for the good by putting the obvious question why he himself, such an intelligent thinker, what is more, should remain in this life [366b2-4]. This question starts what seems to be

92 οἰκεῖος τόπος; cf. 365e4.
93 366b1-2.
94 Although perhaps with some Stoic or Philonic 'colouring'.

a second cycle of argumentation, in which what Socrates has so far argued partly recurs and is elaborated on at some lenghth.

Socrates starts the second 'cycle' by giving an exposition which he says is the reproduction of an ἐπίδειξις[95] of the sophist Prodicus. It is a diatribe against life and an illustration, as it were, of the assertion in 365b4-5 that "life is a brief stay in a foreign land" [366b5-369b5]. Indeed, the bewailing of life (*deploratio vitae*) belongs to the *topoi* of consolation-literature.[96] Then in 369b5 there is again an abrupt transition: Socrates proceeds to invalidate the fear of death and the fear of sensation after death, that is, the perceiving of non-sensation. The condition of being dead has nothing fearful about it. There is no need for Axiochus to fear the deprivation of the goods of life, for, once dead, he will not be aware of this deprivation; the goods of life will not be replaced by evils. The argumentation Socrates uses here is Epicurean, although this cannot be explained now, as it was in the similar argumentation in the first 'cycle', as a preparation for a more subtle dualistic viewpoint. Rather, it functions here as a possible explanation of death *alongside* the subsequent discourses on the immortality of *psuchè*. Again, the change is abrupt: without much ado, the part that deals with the immortality of *psuchè* follows close on the heels of the preceding Epicurean argumentation. The second possible explanation of death, that is as a transition (μεταβολή) and a transmigration (μετοίκησις) is divided here into two parts, i.e. into a discourse about the heavens (λόγος οὐράνιος),[97] and into a myth (especially Platonic) that is apparently to serve as a kind of apotheosis. As in the *Apology*, Socrates speaks out in favour of the second possibility of death, though without explicitly *deciding* between the two *logoi*. Axiochus too has been convinced. The two discourses on the immortality of *psuchè* doubtless appeal more strongly to the imagination of one whose mind has to be set at rest with regard to death than does the rationalistic view of Epicurus. This may explain why here, as indeed is generally the case in *consolationes*, preference is given to the second possibility, notwithstanding the fact that both possibilities are found alongside one another in a way that seems to suggest equivalence. In the context of consolation the second possibility is, of course, on psychological grounds, the more promising option.

In several ways the *Axiochus* is a problematic dialogue. Some of the passages which are thought difficult might be explained by assuming that the author, perhaps still a novice or a student, was elaborating an outline for a *consolatio* and found it difficult to connect the parts of such an outline smoothly.[98] However that may be, the little work at least illustrates clearly by its very clumsiness how argumentations of different philosophical origin were used in a more popularizing context; and they were used in this context explicitly for 'psychotherapeutic' pur-

95 i.e. a set speech, declamation.

96 Cf. Johann, *Trauer*, 100 sqq.

97 Cf. 372a11; "this discourse about the heavens" contains elements from Plato, Aristotle, Stoa, and perhaps Philo.

98 Cf. the transitions in 365e2 and 370b1!

poses, that is, to eliminate the fear of death. The genre was popular and apparently satisfied a particular need.

However, one question facing men of all times, the ancient Greeks as well as ourselves, does remain relevant: the question of how far *words* really serve any useful purpose when one is confronted with the most terrifying of all ills: death.[99] Who knows, we too might have a similar experience to that one of the interlocutors in Cicero's *Tusculan Disputations* had when he was reading Plato's *Phaedo*:

> (M.) 'We cannot, can we, surpass Plato in eloquence? Turn over with attention the pages of his book upon the soul [*i.e. the Phaedo*]. You will be conscious of no further need.' (A.) 'I have done so, be sure, and done so many times; but somehow I am sorry to find that I agree while reading, yet when I have laid the book aside and begin to reflect in my own mind upon the immortality of souls, all my previous sense of agreement slips away'.[100]

99 The words "the most ... death" are derived from Epicurus' Letter to Menoeceus, Diogenes Laertius X.125 (see above p.203).

100 Cicero, *Tusculanae Disputationes* I.xi.24 (transl. J.E. King, Cicero, *Tusculan Disputations*, 2nd ed., (London/Cambridge [Mass.], Heinemann/Harvard U.P. 1960 [=1945]). The use of the symbols M and A to indicate the interlocutors in this text is considered to be of Byzantine origin, wherefore I bracketed them. [They indicate respectively the *Magister* and the *Auditor*].

BIBLIOGRAPHY

1 Editions, commentaries and translations

Aetius, "Placita", in *Doxographi Graeci*, ed. H. Diels, 4th ed., (Berlin, W. de Gruyter 1965 [=1879])

Alexander Aphrodisiensis, *De anima liber cum mantissa*, ed. I. Bruns, Supplementum Aristotelicum II.ii, (Berlin, W. de Gruyter 1887)

Aristoteles, *De Anima*, ed. with Transl., Introd. and Notes by R.D. Hicks, (Amsterdam, Hakkert 1965 [=Cambridge 1907])

Ethica Nicomachea, edd. F. Susemihl-O. Apelt, 3rd ed., (Leipzig, Teubner 1912)

Notes on the Nicomachean Ethics of Aristotle, by J.A. Stewart, 2 vols., (Oxford, Oxford U.P. 1892)

De Generatione Animalium, ed. with an Engl. Transl. by A.L. Peck, (London/Cambridge [Mass.], Heinemann/Harvard U.P. 1963 [=1953])

Oeconomica, ed. F. Susemihl, (Leipzig, Teubner 1887)

Parva Naturalia, ed. with Introd. and Comment. by Sir David Ross, (Oxford, Oxford U.P. 1955)

Physica, ed. W.D. Ross, (Oxford, Oxford U.P. 1966 [=1956])

Ars rhetorica, ed. R. Kassel, (Berlin/New York, W. de Gruyter 1976)

Topica, ed. W.D. Ross, (Oxford, Oxford U.P. 1970 [=1958])

Cicero, *Tusculanae Disputationes*, ed. with an Engl. Transl. by J.E. King, 2nd ed., (London/Cambridge [Mass.], Heinemann/Harvard U.P. 1960 [=1945])

Diogenes Laertius, *Lives of Eminent Philosophers*, ed. with an Engl. Transl. by R.D. Hicks, 2 vols., (London/Cambridge [Mass.], Heinemann/Harvard U.P. 1968 [=1925])

Epicurus, *The Extant Remains*, ed. with short Critic. Apparatus, Transl. and Notes by C. Bailey, (Hildesheim/New York, G. Olms Verlag 1975 [=Oxford 1926])

A.A. Long-D.N. Sedley, *The Hellenistic Philosophers*, 2 vols., (Cambridge, Cambridge U.P. 1987)

Lucretius, *De Rerum Natura*, ed. with Proleg., Critic. Apparatus, Transl. and Comment. by C. Bailey, 3 vols., (Oxford, Oxford U.P. 1972 [=1947])

Plato, *Opera*, ed. J. Burnet, 5 vols., (Oxford, Oxford U.P. 1957-61 [=1900-07])

Euthyphro, Apology of Socrates and Crito, ed. with Notes by J. Burnet, (Oxford, Oxford U.P. 1960 [=1924])

Laches, Protagoras, Meno, Euthydemus, ed. with an Engl. Transl. by W.R.M. Lamb, 4th ed., (London/Cambridge [Mass.], Heinemann/Harvard U.P. 1962)

Plato's Phaedo, a Transl. with Introd., Notes and Appendices by R.S. Bluck, (London, Routledge/Kegan Paul 1955)

Plato's Phaedrus, Transl. with Introd. and Comment. by R. Hackforth, 2nd ed., (Cambridge, Cambridge U.P. 1952)

The Republic of Plato, Transl. with Introd. and Notes by F.M. Cornford, (London/Oxford/New York, Oxford U.P. 1974 [=1941])

Pseudo-Plato, Axiochus, ed. with Transl. and Comment. by J.P. Hershbell, Texts and Translations 21/Graeco-Roman Religion Series 6, (Chico, Scholars Press 1981)

Dialogues apocryphes, ed. J. Souilhé, Platon: Oeuvres complètes XIII.3, (Paris, Les Belles Lettres 1962)

Seneca, *Ad Lucilium Epistulae Morales*, ed. L.D. Reynolds, 2 vols., (Oxford, Oxford U.P. 1976-8 [=1965])

Stoicorum Veterum Fragmenta (=SVF), ed. J. von Arnim, 4 vols., (Stuttgart, Teubner 1968 [=Leipzig 1903-24])

Die Fragmente der Vorsokratiker (=DK), edd. H. Diels-W. Kranz, 11th ed., 3 vols., (Zürich/Berlin, Weidmann 1964)

2 General literature

J. Barnes, *The Presocratic Philosophers*, 2nd ed., (Cambridge, Cambridge U.P. 1982)
G. Binder- B. Effe (Hg.), *Tod und Jenseits im Altertum*, Bochumer Altertumswissenschaftliches Colloquium, Band 6, (Trier, Wissenschaftlicher Verlag Trier 1991)
J. Bremmer, *The Early Greek Concept of the Soul*, (Princeton, Princeton U.P. 1983)
W. Burkert, *Greek Religion*, (Oxford, Blackwell 1985)
J. Chevalier, *Étude critique du dialogue pseudo-platonicien l'AXIOCHOS sur la mort et sur l'immortalité de l'âme*, (Paris, F. Alcan 1914)
E.M. Cioran, *De l'inconvénience d'être né*, (Paris, Gallimard 1973)
D.B. Claus, *Toward the Soul*, (New Haven/London, Yale U.P.1981)
I.M. Crombie, *An Examination of Plato's Doctrines*, 3rd ed., 2 vols., (London, Routledge/Kegan Paul 1969-71)
A. Flew, *Body, Mind, and Death*, Problems of Philosophy Series, (New York/London, Colliers/ MacMillan 1964)
"Immortality", in *The Encyclopedia of Philosophy*, ed. P. Edwards, 8 vols., (New York/London, MacMillan 1972 [=1967]), IV: 139-50
Ch.L. Griswold, *Self-Knowledge in Plato's Phaedrus*, (New Haven/London, Yale U.P. 1986)
W.K.C. Guthrie, *A History of Greek Philosophy*, 6 vols., (Cambridge, Cambridge U.P. 1962-81)
D.E. Hahm, *The Origins of Stoic Cosmology*, (Columbus, Ohio State U.P. 1977)
M. Jager, "Doodsproblematiek", *Lampas* 13.4 (oct. 1980), 278-86
"Antieke filosofen over de dood", *Lampas* 13.4 (oct. 1980), 305-12
H.- Th. Johann, *Trauer und Trost; eine quellen- und strukturanalytischen Untersuchung der philosophischen Trostschriften über den Tod*, Studia et Testimonia Antiqua V, (München, W. Fink Verlag 1968)
R. Kassel, *Untersuchungen zur griechischen und römischen Konsolationsliteratur*, Zetemata 18, (München, C.H. Beck Verlag 1958)
T. Nagel, *Mortal Questions*, (Cambridge, Cambridge U.P. 1992 [=1979])
M. Pohlenz, *Die Stoa, Geschichte einer geistigen Bewegung*, 2nd-3rd ed., 2 vols., (Göttingen, Vandenhoeck & Ruprecht 1959-64)
J.E. Raven, *Plato's Thought in the Making*, (Cambridge, Cambridge U.P. 1965)
J. Rist, *Epicurus: An Introduction*, (Cambridge, Cambridge U.P. 1972)
W.D. Ross, *Aristotle*, 5th ed., (London, Methuen 1971 [=1949])
H. Schadel, ΘΑΝΑΤΟΣ; *Studien zu den Todesvorstellungen der antiken Philosophen und Medizin*, Würzburger medizinhistorische Forschungen, Band 2, (Pattense, H. Wellm Verlag 1974)
G. Scherer, *Das Problem des Todes in der Philosophie*, 2nd ed., Grundzüge, Band 35, (Darmstadt, Wissenschaftliche Buchgesellschaft 1988)
A. Schopenhauer, *Die Welt als Wille und Vorstellung*, ed. W. Frhr. von Löhneysen, 2 vols., (Frankfurt am Main, Suhrkamp 1986)
E. Vermeule, *Aspects of Death in Early Greek Art and Poetry*, Sather Classical Lectures 46, (Berkeley/Los Angeles/London, University of California Press 1979)
G. Vlastos, *Socrates: Ironist and Moral Philosopher*, (Cambridge, Cambridge U.P. 1991)

Resurrection, Revelation and Reason
Husayn Al-Jisr (d. 1909) and Islamic Eschatology

Rudolph Peters

When, during the second half of the nineteenth century, members of the intellectual elite in the Islamic world became familiar with the findings of modern science, men of religion began to tackle the problem of how to relate these findings to the religious truths founded on revelation. The Indian Muslim Sayyid Ahmad Khan (1817–1898) was the first one to occupy himself with this problem. He advocated a form of Islam that was in agreement with the modern natural sciences and that had assimilated many ideas and values that were current in the West at that time.[1] In the Middle East, the Egyptian reformer Muhammad ʿAbduh (1849–1905) worked in the same vein. He intended to interpret Islam in such a manner that it would not be an obstacle to modernization. Like Sayyid Ahmad Khan he posited that there could be no contradiction between revelation and nature. He expressed this notion as follows: "God has sent down two books: one created, which is nature, and one revealed, which is the Kurʾān.[2] These books cannot contradict one another. If there is an apparent contradiction, then the Koran must be interpreted metaphorically (*taʾwīl*) so as to make it agree with the findings of natural science."[3]

In this intellectual tradition the Syrian scholar Husayn al-Jisr (1845–1909)[4] had his place. Although he is now almost forgotten, he enjoyed great popularity during his lifetime and for some time after his death.[5] One of the reasons for his popularity was that in his writings he dealt with the findings of modern science. In this essay I will examine why and to what extent he made use of natural phenomena and rational explanations in his theological discourse, and I

1 On Sayyid Ahmad Khan, see: C.W. Troll, *Sayyid Ahmad Khan: A Reinterpretation of Muslim Theology* (New Delhi, Vikas Publishing House 1978); J.M.S. Baljon, *The Reforms and Religious Ideas of Sir Sayyid Ahmad Khan*, 3rd rev. ed. (Lahore, Ashraf Press 1964).

2 *al-Manār*, vii, 292, as translated and quoted by C. C. Adams, *Islam and Modernism in Egypt* (New York 1933), 136.

3 Adams, *Islam and Modernism*, 127–143.

4 The data about his life are based on Johannes Ebert, *Religion und Reform in der arabischen Provinz: Husayn al-Gisr al-Tarābulusi (1845–1909). Ein islamischer Gelehrter zwischen Tradition und Reform*, Heidelberger Orientalistische Studien 18 (Frankfurt a.M., Peter Lang 1991).

5 As late as in the 1940s, some of his books were popular in Indonesia. See Tāhir al-Djazāʾirī, *De edelgesteenten der geloofsleer. Uit het Ar. vert. en van een inl. en aant. voorzien door G.F. Pijper* (Leiden, Brill 1948) XIV–XV.

will focus on his views on eschatology, as this is pre-eminently the domain of metaphysical speculation based on revealed texts.

Husayn ibn Muhammad al-Jisr al-Tarābulusī was born in Tripoli (now in Lebanon), where he lived most of his life and died. Between 1865 to 1872 he studied at al-Azhar University in Cairo. In this period he developed an interest in modern science and began to study it on his own. Back in his homeland he became a teacher of religion and taught at several schools in Beirut and Tripoli. In 1886 he published *al-Risāla al-Hamīdiyya fī haqīqat al-diyāna al-islāmiyya wa-haqqiyyat al-sharīʿa al-Muhammadiyya* (The Hamidian Treatise on the Truth of the Islamic Religion and the Verity of Mohammed's Law).[6] This book, which was translated into Turkish three years later,[7] won him fame and the reputation of an Islamic scholar abreast of his times. Sultan Abdülhamid II (1876–1909), in pursuance of his policy of patronizing Islamic scholars and most probably flattered by the dedication of the book to him, rewarded the author with decorations and money. In 1891 he invited the author to Istanbul and offered him a stipend to write a second book to defend Islam and refute the objections raised by the materialists. The book was published in 1905[8] in Cairo under the title *Al-husūn al-hamīdiyya li-muhāfazat al-ʿaqāʾid al-Islāmiyya* (The Hamidian Strongholds for Defending the Islamic Doctrines). He stayed in Istanbul nearly one year. After his return to Tripoli he founded a magazine called *Jarīdat Tarābulus*. He died in 1909.

In both the *Risāla* and the *Husūn* the relationship between religion and the findings of modern science plays a role. The difference between the two books, however, is their point of departure. In the *Risāla* al-Jisr addresses in a polemical way the "materialists" (*al-māddiyyūn*, or *al-tabīʿiyyūn al-māddiyyūn al-dahriyyūn*), i.e., in his definition, those who only recognize truth if it is based on sensory perception and reason. Obviously, the intended readership are the modernized Muslims who have acquired some familiarity with Western science. They form, so to speak, the audience listening to the debate between al-Jisr and the materialists, and it is them whom al-Jisr wants to convince. The book focuses on natural phenomena and scientific theories, and the author tries to demonstrate that the findings of modern science, to the extent that they are proven beyond doubt, which, he claims, is often not the case, are not contrary to the principles of Islam and, conversely, that the basic tenets of Islam are not contrary to modern science. In the *Husūn* al-Jisr addresses Muslims whose faith has become weakened by having become acquaintanted with the natural sciences through the press. The book deals chiefly with the established Islamic beliefs, and the author tries to

---

6 Beirut 1305 H., 534 pp. Henceforth referred to as *Husūn*. The adjective Hamidian (*Hamīdī*) derives from the name of the Ottoman sultan Abdülhamid II (1876–1909) and might be translated "dedicated to Abdülhamid".

7 By Manastirli Ismāʿīl Haqqī, 2 vols. Istanbul 1307.

8 1323 H. Henceforth referred to as *Risāla*. Cf. C. Brockelmann, *Geschichte der arabischen Litteratur* (Leiden 1937–1949) Suppl. 2, 776. The book itself does not mention the date of publication.

refute the doubts that have arisen amongst them and to demonstrate that the dogmas and notions of Islam are not in conflict with modern science and that they can be explained rationally.

His rational demonstration of Islamic dogmatics starts from the belief in God's existence and often his argumentation is rather self-evident, resting mainly on the acceptance of God's omnipotence. In this way he defends the belief in Mohammed's Nocturnal Journey and Ascension (*al-Isrāʾ wa-l-Miʿrāj*). According to the established interpretation of the verse "Glory be to Him, who carried His servant by night from the Holy Mosque to the Further Mosque..." (K. 17:1), supported by numerous Traditions, Mohammed travelled one night, riding a horse called al-Burāq, from Mecca to Jerusalem, ascended to heaven and returned to Mecca. Al-Jisr explains that it is rationally possible that bodies move with enormous velocity and that, therefore, it is also rationally possible that God moves the Prophet with such a speed as needed to accomplish the aforementioned journey.[9] Similarly, in explaining the miracle (*muʿjiza*) performed by Moses (Mūsā), when he struck a rock with his stick and the rock produced water,[10] al-Jisr points out that God can create water *ex nihilo* or transform air into water. The latter feat is in accordance with the laws of nature and can be performed even by man, since "this year" scientists for the first time have liquefied air.[11]

With regard to eschatology, al-Jisr closely follows the traditional doctrine found, with small variations, in the major works of theology.[12] This doctrine divides the Afterlife in two stages: an individual one, namely the time between one's death and the Hour or Day of Resurrection, and a collective one, the time after the resurrection of the body.

The first stage begins with death, i.e. when the soul (*rūh*) leaves the body. Every human being has a soul, and it is God's custom that as long as the soul is in the body, one is alive and that one dies as soon as the soul has left it. It is the soul through which one perceives and understands, and through which one experiences pain and pleasure. When a person's appointed time has come, the angel ʿAzrāʾīl takes away his soul, which leaves the body through the mouth. As soon as the dead person has been buried, God restores to him his soul and, therefore, his senses and his reason, but only to the extent that he is able to understand words

---

9 *Husūn*, 114–115.

10 K. 2:60: "And when Moses sought water for his people, so We said, 'Strike with thy staff the rock'; and there gushed forth from it twelve wells; (...)".

11 *Husūn*, 41–42.

12 *Husūn*, pp. 86–89; For the traditional notions, see: Jane Idleman Smith and Yvonne Yazbeck Haddad, *The Islamic Understanding of Death and Resurrection* (Albany, State University of New York Press 1981); Ragnar Eklund, *Life between Death and Resurrection according to Islam* (Uppsala 1941); Hermann Stieglecker, *Die Glaubenslehren des Islam* (Paderborn, Ferdinand Schöningh 1962) 730–808; D. S. Attema, *De Mohammedaansche opvattingen omtrent het tijdstip van den Jongsten Dag en zijn voortekenen,* Diss. Universiteit van Amsterdam (Amsterdam 1942); Ramadān b. Muhammad, *Hāshiya ʿalā sharh al-Taftazānī ʿalā al-ʿAqāʾid li-l-Nasafī* (Istanbul 1315 H.), 221–238; Ibrāhīm al-Bayjūrī, *Tuhfat al-murīd ʿalā jawharat al-tawhīd li-Ibrāhīm al-Laqqānī* (Cairo, Mustafā al-Bābī al-Halabī 1939) 99–118.

spoken to him, to reply to them, and to feel pain and pleasure. While in his grave, a dead person will be roughly interrogated by two angels in order to make his faith and his deeds known to them. Accordingly, he will be punished and made to suffer pain or be rewarded and made to experience pleasure. The pain and pleasure will last until the Day of Resurrection, except that for some sinners the punishment may stop earlier. One of the conditions of the grave is that it squeezes the dead person as the sides draw nearer to each other.

The coming of the second stage, the period after the Hour that will last forever, is heralded by the Signs of the Hour.[13] Among the classical theologians of Islam, there is some difference of opinion regarding the number and the content of these apocalyptic signs, but essential elements are the appearance of the Mahdī and the Dajjāl, the coming of Ya'jūj and Ma'jūj (the Gog and Magog from the Bible), and the descent and final victory of ʿĪsā ibn Maryam (Jesus). The chronology of these events, however, is rather vague. Before the Hour the Mahdī, the Rightly Guided One, will rise and conquer the earth. During his reign justice and prosperity will prevail. He will be challenged by the one-eyed impostor, the Dajjāl, who will acquire some following, although the word *kāfir* (infidel) is written on his brow. Then ʿĪsā will descend from heaven and accept Islam. Either he or the Mahdī – according todifferent traditions – will defeat the Dajjāl. During this period two barbarous and fierce peoples, called Ya'jūj and Ma'jūj, will appear from the extremities of the earth, from behind the dam built by Alexander the Great in order to contain them, and they will wreak havoc on earth.[14] Their mischief will be stopped by the First Blast of the Trumpet. Further signs are the coming of the Beast speaking to the people and blaming them for their unbelief that makes them incapable of recognizing the Signs of the Hour[15] and the spreading of smoke covering the skies.[16] Al-Jisr, following the popular nineteenth century Egyptian theologian and jurist al-Bayjūrī (or: al-Bājūrī) (1783–1860), mentions four more signs: the rising of the sun in the West, the destruction of the Kaʿba by Ethiopians after ʿĪsā's death, the disappearance of the words of the Koran from the copies of the Koran and from the memories of men, and finally, the returning of all inhabitants of the earth to unbelief.[17]

The Resurrection and the events that follow it form the beginning of the second period, a period that will last forever. Upon the First Blast of the Trumpet, delivered by the angel Isrāfīl, all inhabitants of the earth and the heavens die. After a long time, the Second Blast of the Trumpet will resound, at which all the dead are resurrected. They will come out of their graves and be gathered on the Place of Standing (*al-mawqif*) to await judgement. This is an episode full of terrifying occurrences. They have to stand and wait for a very long time, while the sun

---

13 See Stieglecker, *Glaubenslehren,* 740–746; Smith and Haddad, *Death and Resurrection*, 65–70; al-Bayjūrī, *Tuhfa,* 110.
14 Cf K. 18:82–89 and 21:96.
15 Cf K. 27:84.
16 Cf. K. 44:9.
17 *Husūn*, 87; al-Bayjūrī, *Tuhfa*, 110.

has drawn near and stands only a mile above their heads. The heat will make them exude sweat that stinks more than a cadaver. They will be nearly immersed in this malodorous liquid, the height of which varies individually depending on the deeds one has committed.

The Judgement begins with an interrogation by the angels. Those interrogated will not be able to lie since their limbs, their skin and even the earth will testify to what they have done in their lifetime. Prophets, holy men (*awliyāʾ*), Companions of the Prophet and other pious people may intercede for individuals. Then the angels will give each of them in his right hand the book in which his good deeds have been recorded and in his left hand the book in which his evil deeds have been written down. Thereupon, God will settle accounts with them, i.e. He will tell every person about any single deed whether it was good or bad. For each person the books will be weighed in the Scale (*al-Mīzān*).

After the Reckoning, all creatures must proceed over the Bridge (*al-Sīrāt*) which extends over the abyss of the Blaze (*al-Nār*, or *Jahannam*, i.e. Hell) and at the end of which awaits the Garden (*al-Janna*, i.e. Paradise). The Bridge is thinner than a hair and sharper than a sword. Only those who have been allowed to go to the Garden will reach the other side; all others will fall down into the flames. The believers who have been obedient or whose deeds have been forgiven through intercession, will live forever in the Garden, the dwelling of pleasure. The believers who have committed sins will remain in the Blaze, the dwelling of punishment, for some period, after which they, too, are allowed to enter the Garden. The unbelievers will eternally remain in the Blaze. The descriptions of the Garden and the Blaze are abundant in the Koran and very lively. The Garden is represented as a large orchard with fruit trees, little rivers and a big pond. Those who have been saved from the Blaze, sit there eating fruits and drinking wine that causes no intoxication nor headache. Around them there are beautiful girls to give them sexual pleasure. The picture of the Blaze is that of a large fire or an oven into which people are forced by hell's angels. They can only eat the bitter fruits of the Zaqqūm tree and drink stinking infested water.

Al-Jisr deals with a number of these events and phenomena and tries to make them palatable for those who are sceptical with regard to them.[18] Almost all of his explanations are founded on God's omnipotence. Regarding the testifying of the limbs, the skin, and the earth he remarks that nothing opposes the idea that God creates the faculty of speech in these objects. The same goes for the Beast that speaks. As to the Bridge, al-Jisr holds that it is not impossible for God to make people pass over a bridge which is like a razor: thinner than a hair and sharper than a sword. However, following some older authorities, he admits that the description of the Bridge may be a metaphor for the difficulty of passing over it. That the sun will rise in the West, move Eastward until noon and then returns to the West following its normal course is not impossible in his view. For regardless of whether one believes that the sun revolves around the earth or the earth around

18 *Husūn*, 89–99.

the sun, it is God who moves these celestial bodies. That means that it is in His power to stop them or move them in a different direction. The descent of ʿĪsā from heavens is performed by angels carrying him, like during the ascension. Angels have been given the power to rise and descend between heaven and earth. As for the objection that a human person needs air for breathing which is lacking outside the atmosphere (*kurat al-hawā*ʾ), al-Jisr's reply is that this is normally so, but that God can preserve life without it.

Concerning the attacks of Yaʾjūj and Maʾjūj, al-Jisr counters the objections that geographers have explored the earth and have not found a trace of these peoples. For he asserts that not all parts of the earth have been visited, especially not in the Northern fringes of the earth, beyond the mountains of ice and the end of the frozen region and that it is possible that these people live in some remote unexplored region. It is known to those who have knowledge of cartography, he continues, that there are in the Northern regions, farther than Siberia, mountains covered with perennial snow and ice and that nobody in our times can travel there. It is also known that beyond those mountains there are regions extending to the end of the earth. It is not unlikely that this region, which because of its low level has less ice and snow, is inhabited by Yaʾjūj and Maʾjūj. It is possible that in the time of Alexander the Great, which was thousands of years ago, there was a valley through which this region was accessible. In this valley then Alexander built his famous dam. Later the valley was filled up with snow and ice so that no one could pass anymore. However, when the Day of Resurrection approaches, the ice will disappear as a result of meteorological and geological factors, such as earthquakes, thus enabling these peoples to leave their abode and attack the rest of the world.

The crux in the Islamic representation of the Afterlife is the quickening of the dead. There are numerous Koranic verses expounding this theme. Although the Muslim philosophers and some modern Muslim scholars have maintained that resurrection is entirely spiritual and must be understood as the resurrection of the soul, the orthodox view, in accordance with unambiguous Koranic texts,[19] holds that the bodies will be brought to life again. Even early in Islamic history, this belief had to be defended against the attacks of those who doubted the possibility that the same body could be restored to a person resurrected from death. What about a man who had been eaten by an another and whose body had become part of the body of the other? To which one of them will these parts be restored? Another common objection against the dogma of resurrection has to do with the question of how the first body that had perished and the resurrected body could be identical. And if these bodies were not identical, then the resurrected body, that had no relation with the first body, would be rewarded or punished for acts committed by the first body. These objections were countered by claiming that because of His omnipotence God was capable of resurrecting the dead as foretold

---

19 For example K. 36:38–39: "(...) he says, 'Who shall quicken the bones when they are decayed?' Say: 'He shall quicken them who originated them the first time; (...).''

in the Koran. To make it rationally acceptable, it was assumed that a man's body consists of two kinds of matter: original components (*ajzāʾ asliyya*), that remained with a person from his birth to his death, and subordinate components (*ajzāʾ fadliyya*). Since the body of a person can grow fat or skinny during his lifetime, it seemed obvious that not all components of a person's body were necessarily linked to his personality. Only the original components would be restored to life and reunited with the soul. After that subordinate parts, originating from the same or another body, would be added to mould a new body in the shape of the original person.[20]

Al-Jisr elaborates this theory[21], and through it he can explain the events in the grave and the Resurrection. First he defends the existence of a soul against attacks by the materialists, who claim that no one has ever seen the soul leaving the body through its mouth at the moment of death. On the basis of Revelation we have to accept the existence of a soul, although we cannot perceive it with our senses. This is, however, not against reason. The materialists, he asserts, assume the existence of ether and of "microscopic animals" (*hayawānāt mikruskūbiyya*). Why then can they not accept the existence of an invisible soul? From the revealed texts we know that it is alive in itself and does not need anything else to make it alive, that it possesses perception, and that when it settles down in a body, it makes it alive and gives it perception and the other characteristics of life. It can be compared, he says, with the phenomenon magnetism that is transmitted to a piece of iron when one rubs a magnet on it.[22]

As a starting point for his detailed theory, al-Jisr takes the verse "And when thy Lord took from the Children of Adam, from their loins, their seed, and made them testify touching themselves, 'Am I not your Lord?' They said, 'Yes, we testify' (...)" (K. 7:172). According to several Traditions[23] this verse must be interpreted as follows:

> First God took out of Adam's loin his offspring (*dhurriyya*). From these particles (*dharr*), that He had taken from Adam, he took their offspring in particles. Thereupon he took from these last particles their offspring in particles until the last of the human species. Then he created reason, understanding, movement and speech in them and directed the words 'Am I not your Lord?'. Then they all said 'Yes', that is 'Thou art our Lord'. Then He returned all of them to Adam's loins.

---

20 See for example Ramadān b. Muhammad, *Hāshiya*, 227; Muhammad al-Dasūqī, *Hāshiya ʿalā sharh Umm al-Barāhīn li-Muhammad b. Yūsuf al-Sanūsī* (Cairo, Mustafā al-Bābī al-Halabī 1939) 220. In the latter work the original components are identified as bones and muscles and the subordinate components as fat.

21 *Risāla*, 341–365; *Husūn*, 89–91.

22 *Risāla*, 352.

23 For the Traditions regarding the interpretation of this verse, see Muhammad b. ʿAlī al-Shawkānī, *Fath al-Qadīr*, 2nd impr. (Cairo 1964) II, 262–264.

Al-Jisr equates these particles (*dharr,* sg. *dharra*), which he represents as invisible to the naked eye but already endowed with organs, with the notion of original components and infers from this verse and its traditional interpretation that God first created the original components or particles of all individuals of the human species to exist until the end of time. Then He took Adam's particle, created his body out of subordinate components, after which He put the particles of all future humanity in his loins. This is not implausible, since, according to the scientists, one little drop of water can contain as many "microscopic animals" as there are men on earth. Finally, He placed Adam's soul in his particle located in his body. To this the Koran refers with the words: "When I have shaped him [Adam] and breathed My spirit in him, (...)" (K. 15:29). From the particle, life spread to the rest of the body. He regards the heart as the most plausible centre for the soul and the particle. Physiologists, he claims, agree that the force produced by the contractions of the heart is sufficient for the circulation of the blood. However, they are in disagreement about the cause of the cardiac movement. The latest explanation seems to be that this cause is situated in the heart itself, viz. in neural nodes (*ʿuqad ʿasabiyya*), but they still cannot explain why these nodes produce a regular, discontinuous activity and not a continuous activity. Al-Jisr asserts that it is probable that the heart is the seat of the human particle and that when the soul settles down in the heart, it makes it alive and causes this regular movement that produces the circulation of the blood. Via this circulation, life spreads through the body.

In order to be able to address them, as mentioned in the Koran, God took all the particles out of Adam's loins and gave them their souls. After they acknowledged him, as the Koran tells us, He took away the souls to store them somewhere in the universe and placed back the particles in Adam's loins, probably through the pores of his skin, just like germs may enter and leave the human body, as the scientists claim.

These particles left Adam's body during intercourse, carried by spermatozoa.[24] Or, to put it more precisely, the particle of the son to be begotten and the particles of all future offspring of this son were transferred during intercourse to the seed (*bizra*) produced by his wife's ovary. After the son's particle has been transferred to the seed, God created its body (*haykal*) from this seed and the seminal fluid. These contained, of course, only subordinate components. The particles of the offspring were placed in the loins. At a certain moment God attached the soul to the particle, after which it became alive. This happened in the same manner for all of Adam's sons, and then for their sons and so forth until the end of this world.

---

24 Al–Jisr gives a full and for the most part accurate description of spermatozoids, mentioning their size (1/500 to 1/600 inch), their way of propelling themselves and their speed (1 inch per 13 minutes) and the duration of their activity (seven or eight days within the female body, 24 hours outside). *Risāla*, 356.

After death, the particle is stored just like gold between the layers of the earth to protect it against decay and decomposition. If the deceased had been eaten by an animal or another human being, his particle is temporarily kept among the subordinate parts of that animal or person, but is not affected and will ultimately, after the host's demise, be stored like the other particles. At any time God can bring persons to life by attaching their souls, stored somewhere in the universe, to their particles, kept in the earth and giving it a body consisting of subordinate components. This happens immediately after death in the grave and then during the Resurrection.

Since the particle together with the soul is in fact a complete human being, the only role of the female "seed" is to provide the subordinate components of the child's body. Al-Jisr thus rejects the view, which he ascribes to the materialists, that the *bizra*, i.e. the element produced by the woman's ovary, is the most important part in procreation and that the semen, produced by the male, serves only for fecundation (*talqīh*). Although he does not express it, it is clear that when daughters are begotten, no particles of their future offspring are transferred.

In order to find out to what end al-Jisr employed rational discourse and the findings of the natural sciences, it is useful to contrast him with his contemporary and friend, the Egyptian Muhammad ʿAbduh. They both had dabbled at science and used it in their writings, but their knowledge was unsystematic and lacunose.[25] As regards the relationship between revelation and the findings of modern science, they formulated the same maxim: if there seems to be a contradiction between a revealed text and nature, the revealed text must be interpreted metaphorically.[26] However, the way they applied this maxim differed considerably.

ʿAbduh repeatedly emphasizes that Islam is pre-eminently a religion of reason. He has an open mind towards scientific discoveries and tries to use them in his Koran exegesis and theological writings. His ambition is to show that the seemingly supernatural phenomena mentioned in the Koran can also be interpreted in different ways, not contradicting the laws of nature. His explanation of the word *jinn*, that occurs several times in the Koran and is traditionally taken to mean invisible spirits, is a case in point. Trying to find a more natural interpretation, he suggests that the word might refer to the microbes that cause diseases. His aim in general is to create a new interpretation of Islam more in harmony with the requirements of the modern age and forming no obstacle to reform.

Al-Jisr's attitude, however, was different. His ambition was not the reform and renewal of Islam, but the defence of its traditional truths in the face of doubts resulting from the spread of scientific knowledge. Although he had been sufficiently interested in science to have acquired some knowledge about it, he did not have an open mind towards it. Whenever he finds an apparent contradiction between the Koran and scientific discoveries or theories, his first reaction is to

25 For ʿAbduh, see Adams, *Islam and Modernism*, 136–137.
26 *Husūn*, 100; *Risāla*, 285–6.

examine how conclusively the latter were proven. And often, in his view, the evidence does not bear the test: as we have seen, he did not even unequivocally accept that the earth revolves around the sun. As a result of this attitude, he only seldom was compelled to abandon traditional interpretations of the Koran. The contrast with ʿAbduh is clearly illustrated in al-Jisr's view on the meaning of the word *jinn*. Without reservation he defends the traditional interpretation that they are invisible beings created from a fine and transparent matter, who are able to assume different shapes and to perform difficult tasks andare obliged to follow Islamic law. He tries to demonstrate that their existence is rationally possible as their matter may be similar to ether or air and arranged by God in such a way as to confer the above-mentioned faculties to them. Alternatively, their invisibility may be explained by the fact that God, who has created in every man the faculty of vision, has not created in us the faculty of seeing them.[27]

His conservatism is also evident from his eschatological views. It is indicative that he closely follows the commentary of the nineteenth century scholar al-Bayjūrī, whose writings on theology and law represent the orthodox tradition of learning. Al-Jisr is at ease with the supernatural phenomena that are part of the traditional eschatology. For he sticks to the classical Islamic dogma that there are no laws of nature, but only "customs" of God in creating a series of discrete events in a certain order, and that He is free not to follow these customs, in which case a miracle occurs.[28] Here, too, his ideas differ considerably from ʿAbduh's position. The latter wrote only in very abstract and general terms on eschatology and accepted the existence of laws of nature, since, in his view, God's wisdom makes Him follow these laws in creating events, so as to maintain order in the Universe.

Al-Jisr's conservative attitude is very much in evidence in his theory on the composition of the human body and human procreation, which is no more than an elaboration of traditional notions developed by the theologians in response to logical objections against the resurrection of the body. He bolstered this theory by a very eclectic use of the natural sciences. Science and rational argumentation for al-Jisr are not sources of inspiration for reform and renewal but are subordinate to the traditional dogmas and only serve to make them palatable to new generations of Muslims who have become familiar with scientific theories and terminology. In the last instance his explanations are not rooted in scientific theories, but in the religious dogma of God's omnipotence. That he only propagated traditional religious truths and did not offer a really innovative view of Islam may explain the fact that his popularity was short-lived and that now he is almost forgotten.

---

27 *Husūn,* 112.

28 For the orthodox concept of "God's custom" (*sunnat Allāh*), see Stieglecker, *Die Glaubenslehren des Islam*, 161–162.

forgotten.

BIBLIOGRAPHY

C. C. Adams, *Islam and Modernism in Egypt* (New York 1933)C.W.

D. S. Attema, *De Mohammedaansche opvattingen omtrent het tijdstip van den Jongsten Dag en zijn voortekenen,* Diss. Universiteit van Amsterdam (Amsterdam 1942)

J.M.S. Baljon, *The Reforms and Religious Ideas of Sir Sayyid Ahmad Khan*, 3rd rev. ed. (Lahore, Ashraf Press 1964).

Ibrāhīm al-Bayjūrī, *Tuhfat al-murīd ʿalā jawharat al-tawhīd li-Ibrāhīm al-Laqqānī* (Cairo, Mustafā al-Bābī al-Halabī 1939)

C. Brockelmann, *Geschichte der arabischen Litteratur*, 2 vols, 3 suppl. vols. (Leiden 1937–1949)

Muhammad al-Dasūqī, *Hāshiya ʿalā sharh Umm al-Barāhīn li-Muhammad b. Yūsuf al-Sanūsī* (Cairo, Mustafā al-Bābī al-Halabī 1939)

Tāhir al-Djazāʾirī, *De edelgesteenten der geloofsleer. Uit het Ar. vert. en van een inl. en aant. voorzien door G.F. Pijper* (Leiden, Brill 1948)

Johannes Ebert, *Religion und Reform in der arabischen Provinz: Husayn al-Gisr al-Tarābulusi (1845–1909). Ein islamischer Gelehrter zwischen Tradition und Reform,* Heidelberger Orientalistische Studien 18 (Frankfurt a.M., Peter Lang 1991).

Ragnar Eklund, *Life between Death and Resurrection according to Islam* (Uppsala 1941)

Husayn al-Jisr, *al-Risāla al-Hamīdiyya fī haqīqat al-diyāna al-islāmiyya wa-haqqiyyat al-sharīʿa al-Muhammadiyya (Beirut 1305H)*

Id. *Al-husūn al-hamīdiyya li-muhāfazat al-ʿaqāʾid al-Islāmiyya* (Cairo 1905)

Ramadān b. Muhammad, *Hāshiya ʿalā sharh al-Taftazānī ʿalā al-ʿAqāʾid li-l-Nasafī* (Istanbul 1315 H.)

Muhammad b. ʿAlī al-Shawkānī, *Fath al-Qadīr,* 2nd impr. (Cairo 1964)

Jane Idleman Smith and Yvonne Yazbeck Haddad, *The Islamic Understanding of Death and Resurrection* (Albany, State University of New York Press 1981)

Hermann Stieglecker, *Die Glaubenslehren des Islam* (Paderborn, Ferdinand Schöningh 1962)

Troll, *Sayyid Ahmad Khan: A Reinterpretation of Muslim Theology* (New Delhi, Vikas Publishing House 1978);

ARCHAEOLOGICAL

Iron Age Cinerary Urns From Latium in the Shape of a Hut: Indicators of Status?

Marijke Gnade

A few years ago the Allard Pierson Museum in Amsterdam acquired an Iron Age cinerary urn in the shape of a hut (Figs. 1–4).[1] The huturn is handmade of impurified clay, so-called *impasto*, and varies in color from light-brown to dark-brown with greyish-black patches. Its shape is simple.[2] It has a rectangular base with slightly convex sides, flaring walls and a roof with four slanting sides. The head-beams of the construction have been emphasized by ridges; these end on both sides of the ridgepole in moon-shaped elements. The entrance to the hut is formed by a quadrangular opening in the centre of the frontside which can be closed from the outside by means of a bronze bar. This bar could be stuck through perforated protuberances placed in the centre and on both sides of a quadrangular-shaped door.

Fig. 1

Impasto huturn, APM 12.000

Fig. 2

Impasto huturn, frontside APM 12.000

---

1 Inv. no. APM 12.000. The urn was acquired in March 1990 from a Swiss dealer in ancient art. The photographs of the huturn (Figs. 1-3) have been made by M. Bootsman; G. Strietman made the drawings (Fig. 4).

2 Base 20.5 x 18.5 cm; max. H. 30.0 cm; max. L. 26.2 cm; max. W. 23.0. Base intact, walls and roof have been broken into many fragments; restored.

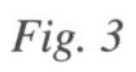

Fig. 3

Impasto huturn, backside APM 12.000

Fig. 4

Section, front- and side-view of APM 12.000, scale 1:4

The urn's surface has been smoothened, as is shown by horizontal and vertical strokes on the walls and the roof. The walls, door and roof were originally decorated by lead strips, of which clear traces have remained: the roof may have been decorated by a pattern of horizontal meanders; above and beneath the entrance a frame of shark-teeth can be recognized, whereas a swastika-motive is still visible below left and on top of the door.

The APM-urn belongs to an exceptional category of cinerary urns which imitate the contemporary dwellings of the living. Only some 200 examples are known. They all come from cremation tombs in Central Italy and date from the period between the Late Bronze Age and the Early Iron Age (10th to the beginning of the 8th century B.C.), when cremation was the main burial rite in large parts of Italy. The necropoleis which have rendered huturns are mainly situated in northern Latium and southern Etruria, with the one large exception of the necropolis of Poggio alla Guardia, near Vetulonia, in northern Etruria.[3] The largest number of huturns (40) comes from this necropolis. Other important find-pots of huturns are the Alban hills (33), Osteria dell'Osa (16) and Rome (6) in Latium, and Bisenzio (16), Vulci (10) and Monte Tosto (24) in Etruria.[4]

The use of the huturns is exclusively funerary: they served in the first place as containers for the cremated remains of men, women or children. Apart from this practical function as ossuary, the huturns probably also had a ritual significance which may have been related to the status of the deceased. This theory had already been put forward in the 19th century when the first huturns were discovered and was mainly based on the urns' rare occurrence and on their exceptional shape compared with the common biconical ones.[5] It has been *communis opinio* since then.

However, the identification of the huturn as an indicator of status is mainly based on the object of the huturn itself. Many pieces were unearthed by chance in a period when comparatively few systematic excavations were carried out.[6] As a result, there is no information on their find circumstances, whereas many huturns

3 The necropolis of Poggio alla Guardia was excavated by Isidoro Falchi from 1881 until 1890, I. Falchi, *Vetulonia e la sua necropoli antichissima*, (Firenze 1891).

4 G. Bartoloni, *e.a.*, *Le urne a capanna rinvenute in Italia*, (Rome 1987). Smaller numbers of huturns have been found in Veii (10), Tarquinia (7), Cerveteri (2), Allumiere (1) in Etruria, Pratica di Mare (2) in Latium; Campo Reatino (1) in Sabine territory; Pontecagnano (1) in Campania.

5 G. Ghirardini, in *Notizie degli Scavi* 1881, 353.

6 The first huturns were discovered in 1816-1817 near Castel Gandolfo, in the Alban hills, east of Rome. Unfortunately, the graves and their outfits were scarcely documented, and the finds (often only the intact ones) were gathered and mostly separately sold. The majority of the huturns from the Alban hills are now spread over musea in Europe, often out of context. For the excavation and the tombs see, P.G. Gierow, *The Iron Age Culture of Latium*, II, *Excavations and Finds: 1. The Alban Hills* (Lund 1964); *Civiltà del Lazio primitivo*, exh. cat. Rome 1976 (Rome 1976), 68-98; cf. also G. Bartoloni, "Le urne a capanna: ancora sulle prime scoperte nei Colli Albani", in *Italian Iron Age Artefacts in the British Museum. Papers of the Sixth British Museum Classical Colloquium*, ed. J. Swaddling (London 1986), 235-241; G. Bartoloni in Bartoloni, *Le urne a*

have been separated from their original contexts. The same is true for the APM-urn. Neither the cremated remains of the deceased nor the burial gifts which must have accompanied the urn have been preserved.

Fortunately, systematic excavations and modern research methods have led to a more thorough knowledge of the huturns' significance in the burial rites. The predominantly aesthetic appraisal of the isolated huturns has gradually shifted to a more varied reflection on this category. Different aspects like funerary ritual, social ideology, ceramic production, family tradition and, of course, the architecture of real huts have been taken into account. Thanks to the excavation of the Iron Age necropolis of Osteria dell'Osa in Latium, an undisturbed complex of early graves – some containing a huturn – could be studied in detail. The composition of the funerary outfits and the spatial distribution of the graves containing a huturn could be compared to contemporary graves in the same necropolis without a huturn, in order to define its significance in the ritual.[7]

Furthermore, a recently published monograph provides us with a typological classification of the Italian huturns in an attempt to order them by region and even by site.[8] The volume contains an extensive catalogue of 198 extant specimens which are presented in about 60 cases with context and find circumstances. On the basis of this important study, it is possible to classify the huturn of the Allard Pierson Museum.

THE CREMATION RITE IN THE 10TH AND 9TH CENTURY B.C.

Our picture of the burial rites in the earliest phases of the protohistoric cemeteries is fairly clear, especially for Latium. Here, the early necropoleis consisted of small numbers of graves, usually not more than 20-30. The graves are mostly concentrated in groups not far apart from each other in correspondence with the settlement pattern in this period, which is characterized by groups of small villages distributed at a short distance from one another over limited territories.[9]

---

7 The necropolis of Osteria dell'Osa has recently been published by A.M. Bietti Sestieri (ed.), *La necropoli di Osteria dell'Osa*, (Roma 1992). Unfortunately, the publication was not yet available at the time this article was written. The necropolis of Osteria dell'Osa is situated near the Latin site of Gabii and contains 600 cremation and inhumation burials dating between the 9th and the beginning of the 6th century B.C. During the 15 years of excavation various aspects of the necropolis have been extensively illustrated, starting in 1979 with the exhibition cataloque *Ricerca su una comunità del Lazio protostorico. Il sepolcreto dell'Osteria del Osa sulla via Prenestina*. This publication was followed by many articles, two of which deal with the early cremation graves with huturns: A.M. Bietti Sestieri, "The Iron Age necropolis of Osteria dell'Osa, Rome: evidence of social change in Lazio in the 8th century B.C.", in *Papers in Italian Archaeology IV. The Cambridge Conference – BAR* 245 (1985): 111-144; *eadem*, in Bartoloni, *Le urne a capanna*, 188-196.

8 Bartoloni, *Le urne a capanna*. Earlier publications on this subject are W.R. Bryan, *Italic Hut Urns and Hut Urn Cemeteries*, (Rome 1925) and J. Sundwall, *Die italischen Hüttenurnen*, (Abö 1925).

9 For a clear survey of the burial rites in the earliest phases of the Latial culture, see AA.VV., "La formazione della città nel Lazio", *Dialoghi di archeologia*, n.s. 2 (1980): 2-232 and G. Colonna, "I Latini e gli altri popoli del Lazio", in *Italia omnium terrarum alumna*, (Rome 1988), 411-528.

In Etruria, the picture seems more or less the same for the 10th century, but already at the beginning of the 9th century B.C. the size of the cemeteries greatly increases in association with settlements of similarly large dimensions.[10]

The cremation graves are formed by a simple cylindrical shaft or pit cut into the ground (*pozzo*-grave) and are generally covered by a stone slab. Sometimes the sides of the shaft are reinforced by small stones which also may form a kind of vault. The cremated remains of the deceased were put into a cinerary urn, which together with the grave gifts was often protected by a stone container (*custodia*) or a large storage jar (*ziro* in Etruria, *dolium* in Latium) (Figs. 5-6). These were placed on the bottom of the grave or in a deepening in the bottom, a *pozzetto*. In a single case the cinerary urn and grave gifts were deposited in a niche in the wall, a *loculus*.[11] The grave was filled up with earth and stones, in some cases containing the remains of the pyre, and closed. The dead

Fig. 5

Pozzo-grave during excavation. Dolium and its contents in situ. Tomb Y, Necropolis Sacra Via (From Gjerstad, The Tombs, fig. 72)

---

10 *Cf.* the Villanovan cemetery of Quattro Fontanili at Veii with more than 100 cremations from the 9th century, J. Toms, "The relative chronology of the Villanovan cemetery of Quattro Fontanili at Veii", *Archeologia e Storia Antica* 8 (1988): 44-97.

11 *Pozzo*-grave with *loculus*: Tomb Q, Sacra Via necropolis, Rome, E. Gjerstad, *Early Rome*, II, *The Tombs*, (Lund 1956), 26-32, figs., 15-19; *Lazio primitivo*, cat. 22, 110-112, tav. XIII,B.

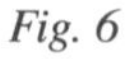
Fig. 6

Section of a pozzetto and dolium with finds in situ. Tomb Y, Necropolis Sacra Via (From Gjerstad, The Tombs, fig. 73)

were usually accompanied by a modest amount of burial gifts: one or more personal objects were mostly put in the urn, and some vases (generally not more than seven or eight) were placed on top of and around the urn.

Although the burial customs for this period are more or less comparable, each region or even each site has its own characteristics. As we have seen, huturns are known to have occurred only in a small area and in a few necropoleis.[12] The rest of cremating Italy made use of the usual biconical cinerary urn (outside Latium) or a simple jar covered with a bowl.

In Latium we find other specific regional characteristics. For instance, the large domestic jars (*dolia*), which were used to protect the grave goods, are documented only here.[13] Also typical for Latium in this early period and closely connected with the use of huturns is the miniature size of the burial gifts. In cremation graves of the 10th and first half of the 9th century B.C., both the personal objects and the vases are very small representations of objects for daily use.

---

12 In only 2 cases are huturns known from outside Latium or Etruria: in Campania, from the necropolis of Pontecagnano: B. d'Agostino, *Studi Etruschi* 42 (1974): 509, tav. LXXXIII/b; Bartoloni, *Le urne a capanna*, 109; in Sabine territory, from Campo Reatino: G. Filippi, "La necropoli di Campo Reatino, I, I materiali", *Archeologia Classica* 35 (1983): 138-167; Bartoloni, *Le urne a capanna*, 74.

13 Colonna, I Latini, 428.

A few burials even contain a terracotta idol, perhaps a representation of the deceased.[14]

If we take a closer look at the funerary outfit, we see a variety of miniature vases which probably had a function in the banquet:[15] in the oldest graves table pottery such as wide open bowls for eating, cups for drinking and plain jars with the function of containers. Furthermore, we see boat-shaped vases with handles placed in the centre on the bottom which may be lamps and round plates supported by tripod stands resembling wooden tables. At the end of the 10th or in the beginning of the 9th century, this repertoire is enlarged with jars with reticulate ornament in relief. These vases occur mostly in pairs (sometimes treble) and were probably meant for the conservation of special foods such as honey. In this period we also encounter the so-called 'calefattoi', braziers or miniature stands for vases used at the banquet and large containers for liquids, either separate or incorporated in the stands (Figs. 7-8).

Apart from the pottery, miniature personal objects were given to the deceased, such as bronze fibulae, bronze razors and knives and bronze weapons such as lances and swords. One grave even contained a complete armour consisting of a spear, a short sword with sheath, a shield and greaves.[16]

Fig. 7

Contents of a cremation burial. Tomb C, Necropolis Sacra Via (From Gjerstad, The Tombs, fig. 48)

14 *Dialoghi di archeologia* (1980): 62-63.

15 *Dialoghi di archeologia* (1980): 51-54, figs. 1-5; Colonna, I Latini, 428-429.

16 Grave 21 in Pratica di Mare (Lavinium), *Lazio primitivo*, cat. 94, 294-296, tavv. LXXIV/B, LXXV/A .

Fig. 8

Contents of a cremation burial. Tomb Y, Necropolis Sacra Via (From Gjerstad, The Tombs, fig. 74)

The reduction in size of the funerary outfit is an essential element of the earliest graves in Latium and seems strictly related to the cremation rite. It occurs systematically – also in graves in which the cinerary urn is a plain jar – until the 9th century, when the burial rite of inhumation is introduced. With the disuse of miniaturization, the huturn in Latium also disappears.[17]

According to a generally accepted interpretation, the use of miniature objects especially manufactured for the grave had a magic-religious meaning. As if to compensate for the destruction of the body of the deceased, the most essential elements of his life, such as his house, his pottery, his weapons and his personal ornaments, and sometimes even his image, were represented in symbolic

17 With the exception of the huturn in grave GG in the Sacra Via necropolis in Rome, cf. Gjerstad, *The Tombs*, 104-108, figs. 33-34, 101-105; *Lazio primitivo*, cat. 25, 114-115, tavv. XIV,B; XV,A. In Etruria, huturns have been documented until the middle of the 8th century B.C., Bartoloni, *Le urne a capanna*, 222.

substitutes intended for the hereafter.[18] Small portions of food occasionally found in or on top of these miniature vases support the hypothesis of their presumed use in the life to come. In several graves remains of meat, fish, grains and even vegetables have been preserved.[19]

As opposed to the ritual of inhumation, the cremated obviously passed to another material state. For this case he was provided with the elements necessary for his new life.[20]

THE HUTURN

An important element in the ritual described above is formed by the representation in miniature of the dead man's house. The question must be raised whether these representations in fact resemble the real houses of this early period or are merely symbolic substitutes. It is striking that no huturn is exactly alike and that many elements of the superstructure are reproduced extremely realistically. Although in general the structural head-elements are similar, there is much difference between the secondary structural elements and the decorative details. Circular, oval and rectangular plans occur, whereas walls may be straight, tapering or flaring. Some roofs are small and simple, showing only a few, short rafters; other roofs are large and strongly overhanging with clearly profiled, long rafters which protrude above the ridge-pole in bird-like extremities. The structure of the walls varies as well. Apart from the simple huts with plain walls and an entrance in the short side, there are many huts with windows and even a second entrance indicated in the long wall.

Unfortunately, the above-described features can only be partially compared with reality since no original huts have been preserved, as they were made of perishable materials. What we know of the early habitation is mainly based on traces in the ground.[21] These consist for the 8th and 7th century in Latium of post-holes, foundation ditches or discolorations of the bottoms of huts in the virgin soil. Entrances are indicated by a step or by the holes for the doorposts. Often, the presence of a hearth is indicated by a small circular concavity, sometimes with the remains of ashes and charred wood. We know even less of the original walls. Only lumps of baked clay with imprints of twigs and reed have been preserved.

On the basis of these finds we know that the Iron Age dwellings had circular, elliptical, square and rectangular plans and that they were made of beams

---

18 About the phenomenon and the significance of the miniaturization of the grave outfit, see G. Colonna, "Preistoria e protostoria di Roma e del Lazio", in *Popoli e Civiltà dell'Italia antica* (Rome 1974), II: 290-292.

19 For Rome see H. Helbaek, "Vegetables in the funeral meals of pre-urban Rome", Appendix 1 in Gjerstad, *The Tombs*, 287-294; N. G. Gejvall, C.-H. Hjorstjö, "Anthropological and osteological investigations on skeletons and bones", Appendix 2 in Gjerstad, *The Tombs*, 295-320; for Osteria dell'Osa, Bietti Sestieri, *Ricerca*, 107; *Dialoghi di archeologia* (1980): 16, 41-42, 44, 50.

20 Bietti Sestieri, *Ricerca*, 105.

21 A complete bibliography on the finds of Iron Age huts in Etruria and Latium is given in Bartoloni, *Le urne a capanna*, 135 note 2.

between which was an interweaving of twigs, plastered with clay. The roofs probably had slanting sides and were also made of twigs and reed. Entrances were mostly situated in one of the short sides. However, for a reconstruction of the additional structural and decorative details of the superstructure such as windows, doors, roofbeams, rafters and smokeholes, we still depend on the hutmodels from the graves. They are our most important source of information.

So far we have not answered the question as to the realism of the huturns. The earlier mentioned publication of the Italian huturns may give an answer.[22] On the basis of a typological classification, the extant specimens of huturns have been divided into groups that correspond broadly to the geographical divisions between Latium and Etruria. Within the latter region it is also possible to make a division between the coastal and inland sites, whereas the pieces from Latium show more common resemblances on the whole.[23] The huturns from Latium are generally smaller (0.30 cm or less) than those from Etruria. They often have a circular plan and are mostly closed from the outside by means of a metal bar stuck through protruberances on and next to the door. Furthermore, in Latium more attention is given to the reproduction of structural elements, like beams along the walls, double doors, side-windows, porticos and side-shelters. The urns from Etruria, on the other hand, are characterized by a greater wealth of decorative detail like birdlike endings of the roofrafters, incised decoration and especially the use of metal strips.

If we compare the typological – regional or local – characteristics of the huturns to the limited remains of the real huts, certain resemblances can be distinguished.[24] Oval and circular plans occur both in Etruria and Latium, but, as with the urns, there seems to exist a preference for circular plans in Latium. Furthermore, the typical entrance-portico of the Latial urn is often documented in the field. Also in size there seems to be a general correspondence between the hutmodels and the real huts. The relatively larger dimensions of the Etruscan urns tally more or less with the excavated remains from Etruria.

On the basis of these general similarities, we may conclude that the huturns from the graves are indeed likely to have been realistic representations of the dwellings of the living. Obviously, the maker of the urn was familiar with the structure of a hut and was probably well informed about the regional taste. However, in the application of the secondary structural elements and the decorative details he may have acted according to the very local 'cultural' traditions of the single family groups (see below) or may even have followed his own personal taste. The consistency in manufacture and decoration of the objects in graves belonging to one 'family', which is observed in some cases, may refer to a similar situation.[25] One may even think of various artisans connected with single families.

---

22 Bartoloni, *Le urne a capanna*.

23 *Eadem*, 123-133.

24 *Eadem*, 140-143.

25 Bietti Sestieri, *Papers in Italian Archaeology*, 122; Colonna, I Latini, 446.

The APM-urn has no clear parallels, and only very few specimens are known with partial resemblances.[26] The Latial area seems to be the most likely place of origin. This attribution is based on the more general characteristics such as the rather small size and the sobre structural design with emphasized structural head-elements. A decisive element is the outside closing of the door. All three aspects are known to be typical of Latium.

However, one aspect of the urn seems to contradict a Latial attribution. The lead strips which once decorated the surface of the urn are typical for the urns from Etruria, whereas this seldom occurs in Latium. So far, this type of decoration is known only from two huturns from Castel Gandolfo, in the Alban hills in Latium.[27] The APM-urn would thus be the third one in Latium.

FOR WHOM WERE THE HUTURNS MEANT?

Although c. 200 urns in the shape of a hut are known, this amount seems rather small when compared with the bulk of cinerary urns with other shapes from the early Iron Age. In many necropoleis the huturn is not even present or covers only a small percentage of the total amount of urns. In the large necropolis of Vetulonia (Poggio alla Guardia), for example, only 40 out of 900 excavated tombs contained a huturn.[28] Obviously, only a few people had the privilege of being buried in a symbolic representation of their house, the making of which must have been a rather costly expenditure. Hence, there can be no doubt that the huturns expressed something of extreme importance for their 'users'. Possibly, the status or rank of the deceased buried in the urn may have been expressed: the huturn may have been a symbol of power or authority.

In most cases it is a difficult task to establish the status or the gender of the 'users' of huturns. Very often, as with the APM-urn, neither the cremated remains nor the grave outfit, our main sources of information, have been preserved. Only when the complete outfit has been preserved can the sex of the deceased be established by the presence of distinct elements of gender such as razors and weapons for men and spindle-whorls and beads for women. Furthermore, some personal ornaments may indicate a distinction in gender: the fibula with serpen-

---

26 From the corpus of huturns only five examples show partial resemblances with the APM-huturn; the first three come from sites in Latium; two pieces are from outside Latium but are seen as typologically related to those from Latium. 1) Marino, Campofattore, tomb 1, Bartoloni, *Le urne a capanna*, 94, cat. 153, Fig. 71, Tav. XXXVII, c-d; 2) Grottaferrata, Villa Cavalletti, tomb IX, Bartoloni, *Le urne a capanna*, 91, cat. 145, Fig. 68, Tav. XXXV, c; 3) Mannheim, Städtisches Reiss-Museum, inv.no. Bg. 18, Bartoloni, *Le urne a capanna*, 116, cat. 192, Fig. 93, Tav. LVI, a-b; 4) Bisenzio, Necropoli di Porto Madonna, tomb II, Bartoloni, *Le urne a capanna*, 50, cat. 61, Fig. 27, Tav. XV, d; 5) Campo Reatino, Bartoloni, *Le urne a capanna*, 74, cat. 118, Fig. 54, Tav. XXVII, a.

27 Bartoloni, *British Museum*; Bartoloni, *Le urne a capanna*, 98, cat. 158, Fig. 73, Tav. XXXIX, c-d; 101, cat. 163, Fig. 78, Tav. XLII,c-d.

28 Bartoloni, *Le urne a capanna*, 9.

tine bow is usually seen as a typical male attribute, whereas the fibula with semi-circular bow is regarded as a female attribute.

Rank or status are far more difficult to establish, particularly for this early period in which the communities are small, probably no more than twenty to thirty persons, and based on a kinship structure. These communities probably consisted of single family groups, the so-called extended families, in which status was not connected with a specific class or personal wealth, but was more probably related to sex and age.[29] The usual indicators of status like expensive grave constructions or costly artefacts fail us in the earliest period of Latium. The graves are more or less uniform, whereas the outfits show coherent compositions without extreme contrasts in wealth. Yet, some social distinction between the individual members of these small communities seems to have existed, as is suggested by the presence in some graves of specific objects which are absent in other graves. The most evident ones are – apart from the huturns – the miniature weapons which have been documented for some graves, and which seem to delineate a specific role for certain men.[30] In many graves the huturn and the weapons occur together. Those graves may have belonged to the elder men of the family or even to the head of the family, the *pater familias*.

Only one find-complex, the necropolis of Osteria dell'Osa, offers the opportunity to establish a correlation between the huturn's presence in the funerary outfit and the status of its owner. The necropolis, which contains about 600 graves, has been systematically excavated and presents a good sample of about 200 graves, both cremation and inhumation, from the 9th century B.C. These graves form the earliest core of the necropolis and are situated in one particular section, the north-western part. The majority consists of inhumations in rectangular *fossa*-graves, only c. 12% are cremations, both in *pozzo*-graves and in *fossa*-graves.[31] In particular, the cremations are of interest to us because of the regularity with which a huturn occurs in a systematic association of grave goods, which may help to identify the social status of the deceased.

At various occasions, the excavator of the necropolis, Anna Maria Bietti Sestieri, asserted that the cremation ritual in *pozzo*-graves in the earliest period of this particular necropolis (first half of the 9th century B.C.) was reserved only for persons occupying important positions in their families and communities.[32] At the same time, the excavator has tried to establish a connection between a huturn's presence and the status of its owner in life by means of an examination of the spatial, ritual and social implications of the outfits of the graves in question. Two separate groups of contemporaneous graves, which have been identified as the North-group (c. 23 graves) and South-group (c. 30 graves) and which probably represent two single family groups, offer a firm basis for her theory.

---

29 Bietti Sestieri, *Ricerca*, 100; *Dialoghi di archeologia* (1980): 64; Colonna, I Latini, 446.
30 *Dialoghi di archeologia* (1980): 63.
31 Bietti Sestieri, *Papers in Italian Archaeology*; *eadem* in Bartoloni, *Le urne a capanna*, 188-196.
32 Bietti Sestieri, *Papers in Italian Archaeology*, 124; *eadem* in Bartoloni, *Le urne a capanna*, 190.

The gender of the cremated deceased in both groups could be established by the skeletal remains and by the often distinctly male objects in the grave outfits. The social identity of the cremated men could be deduced from the systematic combination of specific features, which distinguished the cremations from the inhumations.

Apart from the type of ritual which implies the maximum energy expenditure of the community's members (cremation with miniaturized grave-goods manufactured exclusively for the grave), the cremations are distinguished by their spatial distribution. In both groups, they are situated in a central and more or less isolated position among inhumations which seem to be organized around them.

The distinct social status of the cremated men is further indicated by the composition and the quality of the grave goods. These consist of miniature objects in a systematic association and at the same time contain the highest number of objects which may be considered indicators of prestige and social rank: jars with a redistributive function, weapons (spear and/or sword) which are never present in the contemporaneous inhumations, objects of a probable religious significance such as votive offerings, and often a huturn or a jar covered by a roof-like lid (in eight of the ten cases).

Apart from their almost consistent occurrence in graves of this clearly defined groups, huturns also occur in some other cremations set in small rectangular *fossae*.[33] These were not situated among the central cremations, but had isolated positions. Although the structure of these graves is different and the dolium protecting the grave goods is absent, the ritual and composition of the funerary outfit are identical to those of the central cremations. The huturn's function as an indicator of social status is even more clearly shown in these graves by the exclusive occurrence of a type of huturn which is characterized by a false door and a mobile roof-lid. The deceased in these graves have been identified as young men. Their status may have been that of coming men: those who are destined to become family-heads, but who have not yet reached this position.

The results of the research in the necropolis of Osteria dell'Osa illustrate that the use of huturns in this particular necropolis is connected to the male gender and is part of a ritual that singles out a selected group of persons. Those persons have been identified as the leading men in the family or, possibly, the individual family-heads. Furthermore, some cremations are characterized by the occurrence of an exclusive type of huturn. In these cases, which also show a different grave-structure, it even seems possible to connect a specific type of huturn with a specific age group.

These clear observations, however, cannot lead to a general conclusion about the huturn's significance in the cremation ritual. The male attribution, for example, is less certain in other find-complexes. In both the necropolis of the

33 Bietti Sestieri, *Papers in Italian Archaeology*, 126; *eadem* in Bartoloni, *Le urne a capanna*, 191.

Sacra Via in Rome and the various find-spots in the Alban hills, huturns have also been found in female burials.

In the necropolis of the Sacra Via on the Forum Romanum, 6 of 17 cremation burials contained a huturn: in four cases this was a male burial, in one case a female burial, and in one case a burial of a young girl c. ten years old.[34] The male cremations are more or less homogeneous as to grave structure and composition of the funerary outfit. The female cremations show divergent aspects. The grave of the adult female, for example, had a niche in the wall, in which the cinerary urn and a rich outfit (ten miniature vases, a spindle-whorl, a bronze ring and three amber beads) were placed without the protecting *dolium*.[35] The girl's funerary outfit is also peculiar. It contains a huturn and four vases of normal dimensions, and is rich in personal ornaments such as bone beads, a spindle-whorl, a bronze fibula, a miniature bronze and a miniature golden wire.[36] The huturn in this grave is relatively large and more slender than the other huturns. Its decoration, which consists of various motives in white and red, is also more abundant.

In the Alban hills huturns have been found in c. 60% of the total number of known cremation graves.[37] From what we know of the scarce notes made in the last century, when this area was excavated, the site of Montecucco, Castel Gandolfo, has produced the largest amount of huturns, c. sixteen in ninteen graves. The graves also seem to contain relatively rich outfits.[38] Unfortunately, most huturns from the Alban hills cannot be attributed anymore to their respective graves. However, in the few cases where skeletal remains and grave goods have been preserved, it appears that the urns were used for mostly young men and women.

A completely different picture is offered by the one large find-complex of huturns in northern Etruria, that of Poggio alla Guardia (Vetulonia), one of the many necropoleis in this area.[39] Apart from the absence of miniaturized grave goods, the ritual also differs in other aspects from the one in Latium.

---

34 The necropolis of the Sacra Via on the Forum Romanum consists of two areas of graves which are part of the same necropolis of which 45 tombs have been excavated. They range in date from the 10th to the 8th century B.C., in a unmistakable chronological sequence from west to east. The largest area of tombs is situated between the temple of Antoninus and Faustina and the Sacra Via and was excavated by Giacomo Boni in 1902-1905. The results of the excavations were published in the *Notizie degli Scavi* 1902, 1903, 1905, 1911; and in Gjerstad, *The Tombs*. In 1951-1952 four other graves dating from the 10th century B.C. were discovered between the Arch of August and the temple of Julius Caesar by S.M. Puglisi and R. Gamberini, published in *Bollettino di paletnologia italiana*, n.s. VIII (1951-52): 45 ff.; *Bollettino di paletnologia italiana*, n.s. XI (LXIV) (1954-55): 299 ff.; Gjerstad, *The Tombs*, 86.

35 Female grave Q, Gjerstad, *The Tombs*, 26-32, figs., 15-19; *Lazio primitivo*, cat. 22, 110-112, tav. XIII,B.

36 Child grave GG, Gjerstad, *The Tombs*, 104-108, figs. 33-34, 101-105; *Lazio primitivo*, cat. 25, 114-117, tavv. XIV,B; XV,A; Anna de Santis in Bartoloni, *Le urne a capanna*, 185.

37 See note 6.

38 Bartoloni, *Le urne a capanna*, 202-203.

39 See note 3; M. Cygielman in Bartoloni, *Le urne a capanna*, 147-151.

The cremation graves were occasionally surrounded by large stone circles. The huturns, which have been documented in about forty *pozzo*-graves, seem to have been concentrated in graves situated inside the stone circles. As opposed to the early presence in Latium, the use of huturns in Vetulonia dates from the middle or the end of the 9th century B.C., the period in which in Latium the huturn disappears. In Etruria, however, the urns stay in use until the 8th century. The composition of some preserved grave outfits indicates that men and women were buried in the huturns. The outfits of the graves with huturns do not differ in quantity from those with biconical ossuaries, but often contain prestige objects. This demonstrates that also in Vetulonia the huturns were intended for persons of a certain social distinction. The stone circles clearly indicate the isolation of certain family-groups.

Poggio alla Guardia is the only find-spot of huturns in northern Etruria and unique among many cemeteries in this area without huturns. This may indicate that the huturn was probably an imported object, which attracted the emerging local elite in a period in which intensive cultural and commercial exchange with centres in southern Etruria and the surrounding areas existed.[40]

Although only a few find-complexes of huturns have been presented here, it has hopefully become clear that the factors governing the deposition of huturns in the graves varied from site to site. This was probably due to the variation in funerary ideology of the single families. A similar variation has been pointed out in the local traditions that affected the appearance of the huturns.

However, in spite of these differences, it is also clearly demonstrated that the significance of the huturns in the burial rite is tightly connected to the wish to indicate a specific social distinction of one or a few selected members in a single community. On the grounds of these observations, it seems safe to see the huturns as indicators of excellence.

---

40 Bartoloni, *Le urne a capanna*, 222.

BIBLIOGRAFIE

G. Bartoloni, “Le urne a capanna: ancora sulle prime scoperte nei Colli Albani”, in *Italian Iron age Artefacts in the British Museum. Papers of the Sixth British Museum Classical Colloquium,* ed. J. Swaddling (London 1986), 235–241

G. Bartoloni, F. Buranelli, V. D’Atri, A. De Santis, *Le urne a capanna rinvenute in Italia*, (Rome 1987)

A.M. Bietti Sestieri (ed.), *Ricerca su una comunita del Lazio protostorico, Il sepolcreto dell’Osteria dell’Osa sulla via Prenestina*, (Rome 1979)

A.M. Bietti Sestieri, “The Iron Age necropolis of Osteria dell’Osa, Rome: evidence of social change in Lazio in the 8th Century B.C.”, in *Papers in Italian Archaeology IV. The Cambridge Conference – BAR* 245 (1985): 111–144

G. Colonna, “Preistoria e protostoria di Roma e del Lazio”, in *Popoli e Civilta dell’Italia antica* (Rome 1974), II: 275–346

G. Colonna, “I Latini e gli altri popoli del Lazio”, in *Italia omnium terrarum alumna* (Rome 1988), 411–528

AA.VV., La formazione della citta nel Lazio, *Dialoghi di Archeologia*, n.s. 2 (1980): 2-232

Civilta del Lazio primitivo, exh. cat. Rome 1976 (Rome 1976)

P.G. Gierow, *The Iron Age Culture of Latium*, I, *Classification and Analysis* (Lund 1966)

P.G. Gierow, *The Iron Age Culture of Latium*, II, *Excavations and Finds: 1. The Alban Hills* (Lund 1964)

E. Gjerstad, *Early Rome*, II, *The Tombs* (Lund 1956)

J. Toms, “The Relative chronology of the Villanovan cemetery of Quattro Fontanili at Veii”, *Archeologia e Storia Antica* 8 (1988): 44-97

General Index